Environmental Footprints and Eco-design of Products and Processes

Series Editor

Subramanian Senthilkannan Muthu, Chief Sustainability Officer, Green Story Inc., Tung Chung, Hong Kong

Indexed by Scopus

This series aims to broadly cover all the aspects related to environmental assessment of products, development of environmental and ecological indicators and eco-design of various products and processes. Below are the areas fall under the aims and scope of this series, but not limited to: Environmental Life Cycle Assessment; Social Life Cycle Assessment; Organizational and Product Carbon Footprints; Ecological, Energy and Water Footprints; Life cycle costing; Environmental and sustainable indicators; Environmental impact assessment methods and tools; Eco-design (sustainable design) aspects and tools; Biodegradation studies; Recycling; Solid waste management; Environmental and social audits; Green Purchasing and tools; Product environmental footprints; Environmental management standards and regulations; Eco-labels; Green Claims and green washing; Assessment of sustainability aspects.

Subramanian Senthilkannan Muthu
Editor

Reduction of Industrial Carbon Footprint

Springer

Editor
Subramanian Senthilkannan Muthu
Green Story Inc.
Tung Chung, Hong Kong

ISSN 2345-7651 ISSN 2345-766X (electronic)
Environmental Footprints and Eco-design of Products and Processes
ISBN 978-3-032-10912-5 ISBN 978-3-032-10913-2 (eBook)
https://doi.org/10.1007/978-3-032-10913-2

This Springer imprint is published by the registered company Springer Nature Switzerland AG
The registered company address is: Gewerbestrasse 11, 6330 Cham, Switzerland

This book is dedicated to:

The lotus feet of my beloved Lord Pazhaniandavar

My beloved late Father

My beloved Mother

My beloved Wife Karpagam and Daughters—Anu and Karthika

My beloved Brother

Last but not least
To everyone working in various industrial spaces to make the world, "SUSTAINABLE"

Contents

Contents

Introduction to the Book—Reduction of Industrial Carbon Footprint

Subramanian Senthilkannan Muthu

This book addresses one of the most pressing challenges of the twenty-first century, the mitigation of greenhouse gas emissions from diverse industrial sectors. Industries such as textiles, steel, cement, petrochemicals, construction, transport, and biotechnology contribute significantly to global carbon emissions, both through direct energy consumption and indirect supply chain activities. This book provides a comprehensive overview of strategies, technologies, and practices that enable industries to transition towards low-carbon pathways, including renewable energy integration, circular economy approaches, carbon capture, sustainable materials, and efficiency improvements. By drawing upon cross-sectoral perspectives, it highlights the opportunities, challenges, and innovations required to achieve meaningful decarbonization. This book serves as a valuable resource for academics, policymakers, industry professionals, and sustainability practitioners seeking practical and theoretical insights into reducing the industrial carbon footprint.

The second chapter is about the challenges and opportunities for decarbonizing steel, cement, and petrochemical industries. The industrial sector is a major contributor to global greenhouse gas (GHG) emissions, responsible for nearly 24% of direct anthropogenic emissions in 2019. This percentage include emissions from various Sub-sectors within the broader industrial category, including fossil fuels burned onsite at facilities for energy. Chemical processes, metallurgical and mineral transformation processes. This chapter explores challenges and opportunities in reducing the industrial carbon footprint, focusing on heavy industries like steel, plastics, and cement, which contribute significantly to these emissions. While energy efficiency remains crucial, achieving near-zero emissions requires transformational changes in energy and feedstock sourcing, material efficiency, and circular material flows. This chapter explores strategies such as switching to low- or zero-GHG energy carriers,

S. S. Muthu (✉)
Green Story Incl, Toronto, Canada
e-mail: drsskannanmuthu@gmail.com

S. S. Muthu (ed.), *Reduction of Industrial Carbon Footprint*, Environmental Footprints and Eco-design of Products and Processes,
https://doi.org/10.1007/978-3-032-10913-2_1

carbon capture and utilization (CCU) and carbon capture and storage (CCS). It also discusses the importance of material demand management, material efficiency, and circular economy solutions. This chapter highlights the need for substantial scaling up of infrastructure for electricity, hydrogen, recycling, and CO_2, as well as the phase-out or conversion of existing industrial plants. Finally, policy interventions and technological advancements for achieving significant cuts in industrial GHG emissions are examined.

The third chapter explains from Agroforestry to Pharmaceuticals: Cross-Sectoral Ethnobotanical Approaches for Carbon-Neutral Industries. Ethnobotany plays a significant role in advancing carbon neutrality by combining traditional ecological knowledge with sustainable industrial practices. Indigenous communities, especially in the Western Himalaya, have used agroforestry, rotational farming, and sacred grove preservation to support carbon sequestration, conserve biodiversity, and manage natural resources sustainably. Their climate-resilient crops and wild plant-based food systems also strengthen food security and environmental resilience. Replacing high-emission industrial goods with ethnobotanical alternatives—such as using plant-based dyes instead of synthetic ones, natural fibers for textiles, and biodegradable resins for adhesives—helps lower carbon footprints. Additionally, ethnoveterinary practices like using herbal feed additives can reduce methane emissions and encourage sustainable livestock management. Industries can also use plants to capture carbon by using medicinal and aromatic plants to purify the air, cleaning up pollution with special plants (phytoremediation), and adopting eco-friendly methods to extract plant-based materials while protecting the environment. Traditional healing systems, which include Ayurveda and Traditional Chinese Medicine (TCM), demonstrate sustainable medicinal plant use through regenerative agriculture, ethical harvesting, and eco-friendly processing techniques. To maximize the potential of ethnobotany in carbon neutrality, strong policy agendas and collaborations between policymakers, industries, and indigenous communities are essential. By integrating ethnobotanical principles into industrial strategies, industries can drive low-carbon development, circular economies, and sustainable innovations. The relationship between traditional knowledge and modern sustainability plays a significant role in developing carbon-neutral industries while at the same time preserving cultural heritage and biodiversity.

The fourth chapter is about reducing Carbon Footprint in the Construction Industry: Strategies from Design to Operation. Various sources of carbon emissions of the construction lifecycle such as design, manufacture in the production of the construction materials, transportation, building, operations and maintenance, and lastly deconstruction are presented in the review study. Among the options that have been studied currently on how to reduce carbon footprints at different stages are the use of alternative sources of water, optimizations in the design, recycling construction waste, alternate additives to the materials and the basic enhancements in water and building system efficiency. Research has indicated that other materials, substances, procedures and systems could cut the amount of CO_2 emission by almost 90% at various stages of construction and operation of a building making construction industry to be the largest emitters of carbon of all establishments with minimal space required to cut down their emissions. This is a priceless tool as clients and other stakeholders can use it to make informed judgments on the choice of materials and systems and create environmental awareness on impacts of manufacturing, transporting, and constructing actions.

The fifth chapter analyses the process safety aspects in low carbon hydrogen production via electrolysis. Low carbon hydrogen, produced by water electrolysis using renewable energy, stands out as a crucial energy vector for the decarbonization of industrial and transportation sectors. This chapter analyzes the safety risks associated with green hydrogen production, focusing on alkaline and proton exchange membrane (PEM) electrolysis systems, proposing mitigation measures based on technical standards, industrial best practices, and risk management tools. Through a mixed methodological approach—integrating an integrative literature review, comparative analysis of standards (ABNT NBR IEC 60079-10-1 and Petrobras N-2918), application of HAZOP and Bow Tie, and case studies in pilot units, critical challenges were identified, such as the high flammability of hydrogen, explosion risks from mixing with oxygen, material embrittlement (Hydrogen embrittlement—HE), and leaks in pressurized systems. The results revealed that the Petrobras N-2918 Standard, more conservative than the ABNT, offers practical guidelines for hazardous area classification, reducing risks in semi-confined environments.

The sixth chapter is about the integral use of biotechnology in textiles-An innovative way to reduce carbon footprint. Industrial operations lead to significant greenhouse gas emissions, with carbon dioxide being the major threat. The industrial carbon footprint is the measure of the amount of carbon dioxide released during each process of industrial production. The textile and apparel industry emits a very large amount of carbon through its processes, starting from fibre production to disposal. It has been a perennial problem, and measures are being taken to minimise the carbon footprint. One of the measures taken is the use of biotechnology. Biotechnology has been used at a great level in the textile and apparel industry as it unleashes a plethora of opportunities to reduce the carbon footprint. Biotechnology provides a variety of creative options for lowering industrial carbon footprints, such as carbon capture and utilisation of bioenergy and biofuels. Biotechnology also allows us to create more sustainable related production processes, which reduce the energy usage and greenhouse gas emissions. It can also be utilized to increase the efficiency of industrial processes, hence lowering the wastage and the pollution caused by these activities. This chapter discusses the various biotechnology-based strategies used to minimise the industrial carbon emissions generated by the textile and apparel industry. The application of biotechnology in the areas of fibres and fabrics production, textile finishing using enzymes, sustainable dyeing techniques, and bio-based recycling of textile wastes is discussed in this chapter. This chapter also divulges information on circular economy solutions through biotechnology usage to reduce industrial carbon footprint. This chapter also gives an overview of the current biotechnological measures taken in the textile and apparel industry for lowering the industrial carbon footprint and discusses the future possibilities of their usage.

The seventh chapter is about the technological innovations and climate adaptation strategies for road management. This chapter investigates the impacts of climate change on road transport infrastructure and operations, emphasizing the sector's vulnerability to extreme weather events and long-term climate variability. Adopting a qualitative and exploratory approach based on a narrative literature review, the study identifies and categorizes major climate-related disruptions that affect freight and passenger transportation systems, including physical damage, operational interruptions, safety risks, and logistical inefficiencies. A matrix of climate impacts and associated technological responses was developed to synthesize the findings. The study highlights a range of technological solutions that can enhance the resilience of road transport, such as intelligent transportation systems, climate and traffic monitoring technologies, low-emission vehicles, adaptive infrastructure materials, and logistics management tools. These solutions were further examined in relation to contemporary frameworks, including Nature-Based Solutions (NbS), resilient city initiatives, and systemic risk management approaches. Additionally, the research aligns its analysis with the Sustainable Development Goals (SDGs), emphasizing how climate adaptation in transport contributes to global sustainability agendas.

I take this opportunity to thank all the contributors for their earnest efforts to bring out this book successfully. I am sure readers of this book will find it very useful.

With best wishes,

Dr. Subramanian Senthilkannan Muthu

Challenges and Opportunities for Decarbonizing Steel, Cement, and Petrochemical Industries

Suraj Prakash and Radha

Abstract The industrial sector is a major contributor to global greenhouse gas (GHG) emissions, responsible for nearly 24% of direct anthropogenic emissions in 2019. This percentage include emissions from various sub-sectors within the broader industrial category, including fossil fuels burned on-site at facilities for energy. Chemical processes, metallurgical and mineral transformation processes. This chapter explore challenges and opportunities in reducing the industrial carbon footprint, focusing on heavy industries like steel, plastics, and cement, which contribute significantly to these emissions. While energy efficiency remains crucial, achieving near-zero emissions requires transformational changes in energy and feedstock sourcing, material efficiency, and circular material flows. The chapter explores strategies such as switching to low- or zero-GHG energy carriers, carbon capture and utilization (CCU) and carbon capture and storage (CCS). It also discusses the importance of material demand management, material efficiency, and circular economy solutions. The chapter highlights the need for substantial scaling up of infrastructure for electricity, hydrogen, recycling, and CO_2, as well as the phase-out or conversion of existing industrial plants. Finally, policy interventions and technological advancements for achieving significant cuts in industrial GHG emissions are examined.

Keywords Anthropogenic · Greenhouse gas · Energy efficiency · Circular material · Emissions

S. Prakash · Radha (✉)
School of Biological and Environmental Sciences, Shoolini University, Solan, Himachal Pradesh, India
e-mail: radhuchauhan7002@gmail.com

S. Prakash
e-mail: surajprakash64@shooliniuniversity.com

S. S. Muthu (ed.), *Reduction of Industrial Carbon Footprint*, Environmental Footprints and Eco-design of Products and Processes,
https://doi.org/10.1007/978-3-032-10913-2_2

1 Introduction

Reducing this carbon footprint has therefore emerged as one of the most pressing challenges in the global effort to mitigate climate change. Heavy industries including oil refining, petrochemical production, steel manufacturing, and cement fabrication remain central to contemporary society, yet together they contribute a considerable proportion of greenhouse gas emissions. When both direct and indirect sources are taken into account, these sectors are estimated to be responsible for close to forty percent of the world's total carbon dioxide output [1]. These sectors rely on energy-intensive processes and carbon-intensive feedstocks, making them particularly difficult to decarbonize. Addressing industrial emissions is also essential for achieving the Paris Agreement target of restricting global temperature rise to well below 2 °C [2]. Without substantial mitigation, cumulative industrial emissions could exhaust 15–25% of the remaining carbon budget compatible with 2 °C of warming by 2050 [3]. This challenge is compounded by the anticipated doubling of global material demand by 2060 driven by urbanization and demographic growth in emerging economies underscoring the urgency of decoupling industrial expansion from carbon emissions [4]. The challenges associated with reducing industrial emissions are multifaceted. Technical barriers are among the most significant, as many industrial processes involve chemical reactions and high temperature which inherently produce CO_2. For example, steel production relies on coal both as a heat source and as a reducing agent in chemical processes, while cement manufacturing emits CO_2 during the calcination of limestone [5]. Current low-carbon technologies for these processes, such as hydrogen-based steelmaking or electrified kilns for cement, remain expensive and are not yet widely deployed [6]. Furthermore, industrial assets have long lifetimes, meaning existing infrastructure often locks in carbon-intensive practices for decades.

Heavy industrial sectors often function with narrow profit margins and therefore face considerable financial hurdles when adopting new technological solutions [7]. Shifting industrial operations toward production pathways with reduced carbon intensity often brings higher operating costs, largely because of dependence on renewable electricity and, in some cases, green hydrogen. At the same time, industries exposed to international trade encounter competitive pressures, since firms operating under stricter domestic regulations may struggle against overseas producers subject to weaker environmental standards [8]. In the absence of supportive instruments such as border carbon adjustments or targeted subsidies, these financial pressures may slow or even stall progress. Decarbonizing heavy industries requires extensive infrastructure development for renewable electricity generation, hydrogen production, and carbon capture and storage (CCS) facilities [9]. For instance, electrifying industrial processes demands reliable access to zero-carbon power supplies, yet only 28% of industrial energy was electrified as of 2020 [10].

Although such barriers remain significant, there are promising avenues for cutting emissions from industry through innovative practices. One of the most practical approaches lies in strengthening energy efficiency. Retrofitting older facilities with the most effective technologies has been shown to cut energy intensity by roughly ten to twenty percent, while the introduction of advanced control systems and recovery of waste heat can provide further improvements [10]. Electrification represents another promising avenue; low- and medium-temperature heat processes can already be electrified using commercially available technologies, while emerging solutions like industrial thermal batteries show potential for high-temperature applications [11]. The adoption of low-carbon energy carriers, particularly green hydrogen and bio-derived fuels, presents significant opportunities for transforming industrial processes [12]. Biomass and biofuels provide renewable alternatives for industries where electrification is not feasible, but bringing these energy carriers into industrial systems will depend on easing financial hurdles while at the same time expanding infrastructure on a far broader scale. CCUS has emerged as a promising approach for lowering emissions, particularly in industrial sectors that are challenging to decarbonize [13]. For example, point-source capture units can be integrated into cement kilns or steel plants to confine carbon dioxide before its release into the atmosphere. Captured carbon can be utilized as feedstock for chemical and construction material production or permanently stored in geological formations to prevent its release [14]. Material efficiency and circular economy solutions represent additional opportunities to reduce emissions, measures including lighter product design, extended durability, and higher recycling rates can substantially reduce reliance on virgin materials such as steel and cement [15]. For example, blast furnace gases from steel production can be captured and utilized in chemical manufacturing. This chapter explores the challenges and opportunities to decarbonize heavy metal industries such as steel, cement and petrochemical by providing comprehensive discussion of technological advances, economic implications, infrastructure requirements, and policy frameworks needed to limit industrial greenhouse gas emissions on a large scale. Ultimately, reducing the industrial carbon footprint is central to international climate objectives, while at the same time creating scope for innovation and fostering more resilient production systems.

2 Current Emissions Profile of Heavy Industries

The industrial sector remains a major source of greenhouse gas release, responsible for close to one quarter of direct emissions and approaching two fifths when emissions linked to electricity consumption are taken into account [16]. Within this sector, manufacturing and construction including energy-intensive industries like steel and cement account for approximately 31% of total CO_2 emissions. Plastics production, a subset of chemicals manufacturing, contributes an additional 3.8% of global CO_2 emissions [16]. Other sectors, such as transportation and electricity generation, also play significant roles but are often interlinked with industrial energy consumption. Electricity and heat production remain the single largest source of global emissions at 42%, much of which is consumed by industrial processes [17]. For example, steelmaking alone accounts for approximately 8% of global energy demand and relies heavily on fossil fuels like coal. Cement manufacturing on its own adds close to seven percent of global carbon dioxide output, largely because calcination of limestone inherently liberates carbon dioxide during the process [16].

Sectors such as steel, cement, plastics, and petrochemicals underpin economic growth worldwide, yet they remain among the most carbon intensive activities. Steelmaking illustrates this challenge particularly well, since conventional blast furnace and basic oxygen furnace routes emit between seven and nine percent of total global carbon dioxide, with roughly one and four fifths tonnes of emissions generated for every tonne of crude steel produced, a figure closely tied to the reliance on coal as both reductant and energy provider [18, 19]. While electric arc furnaces (EAFs) using scrap steel reduce emissions to 0.3–0.5 tonnes of CO_2 per tonne, their adoption is constrained by scrap availability, limiting their global share to 25% of production [19]. In 2019 industrial sector accounted for 24% of direct anthropogenic emissions with steel alone responsible for roughly 7–8% of global CO_2 emissions [20, 21]. In steelmaking, emissions originate both from energy use and from chemical reduction of iron ore. The dominant blast furnace basic oxygen furnace route accounting for about 70% of global steel output relies heavily on coal and natural gas for high-temperature heat and as reducing agents, resulting in substantial CO_2 from fuel combustion. Energy-related emissions make up roughly 40–50% of the sector's total, while process emissions chiefly the carbothermic reduction of iron oxide to iron and are difficult to abate without fundamentally changing the production technology [20–22] (Table 1). Cement production adds another 8% [23, 24], while the plastics and petrochemicals industries continue to grow as sources of emissions, driven by rising demand for synthetic materials [25]. Cement manufacturing contributes 8% of global CO_2 emissions, with 1.6 billion tonnes of CO_2 released annually from calcination (chemical decomposition of limestone) and fossil fuel combustion in kilns [26, 27]. Process emissions from calcination alone account for 60–70% of sector's total, complicating decarbonization efforts compared to sectors where emissions are primarily energy-related [26]. Rotary kilns operating at up to 1450 °C burn coal or petroleum coke to produce clinker, accounting for about 40–50% of total emissions. The calcination reaction ($CaCO_3 \rightarrow CaO + CO_2$), which is intrinsic to clinker formation, releases the other half of the industry's CO_2, making process emissions an unavoidable feature of traditional cement production [23, 24].

Plastics and petrochemicals, derived predominantly from fossil feedstocks, emit 3.8% of global CO_2 during production, with lifecycle emissions including end-of-life incineration-exacerbating their impact. The U.S. petrochemical sector, for instance, contributes 5% of national GHG emissions, driven by naphtha cracking facilities that generate 40% of the industry's energy-related emissions [39, 40]. In the plastics and petrochemicals sector, roughly 50–60% of emissions arise from the energy required to crack hydrocarbon feedstocks such as ethylene and propylene in steam crackers, where fossil fuels often supply the necessary heat. The remaining 40–50% stem from the chemical reactions themselves, which liberate CO_2 as a byproduct of hydrocarbon synthesis [25]. As global demand for plastics grows, both energy and process emissions in this sector are projected to rise, underscoring the urgent need for cleaner feedstocks and more efficient production pathways. These sectors face compounding

Table 1 Summary of key findings from studies on CO_2 emissions in heavy industries

Industry and process	CO_2 intensity	Key findings	References
Steel/BFS direct mineralization with hot steam (HS) activation	105.9–123.6 kg CO_2/t BFS sequestered (at 180–220 °C) Mineralization efficiency 29.6–34.9%	HS pre-activation maximizes sequestration efficiency and offers lower cost (~$51.8/t CO_2 at 220 °C) versus many conventional mineralization routes	Xia et al. [28]
Steel (China)/whole industry with hydrogen metallurgy transition (LEAP model)	2020: 2.08 Gt; 2024 peak: 2.17–2.19 Gt; 2050: 0.729–0.828 Gt total CO_2; CO_2 intensity falls to 1.17–1.33 t CO_2/t steel by 2050	Clean H_2 (electrolysis with renewables) accelerates and deepens reductions (extra −48.8 Mt in 2050 vs. baseline); fossil-based H_2 hinders cuts (+50.0 Mt); hydrogen-rich DRI-EAF has lowest process intensity aside from 100% scrap EAF	Chang et al. [29]
Steel/oxy-BF with power-to-gas (SNG) and top-gas recycling (TGR)	Emissions reduction 30–34% versus conventional BF; energy penalty 12.0–17.5 MJ/tHM; cost 216–352 €/t CO_2 avoided	Coupling PtG with TGR achieves substantial abatement while meeting plant heat needs; TGR + SNG lowers cost and energy penalty vs SNG-only	Perpiñán et al. [30]
Cement/clinker production (process-related, Iran)	0.505–0.572 tCO_2/t clinker (EFclinker 2023; mean 0.526)	Process EF tightly clustered; provincial variability; Iran's high clinker factor leads to higher process CO_2 than global datasets	Massoumi Nejad et al. [31]
Cement/full plant system with integrated CaL-DRM (clinker + syngas)	852.3 kgCO_2/t clinker converted to syngas; 62.5 kgCO_2/t clinker additionally captured; flue gas CO_2 cut to 0.24%	Demonstrates deep CO_2 abatement at source via conversion + capture with large syngas co-product; minimizes downstream CO_2 handling	He et al. [32]
Cement/global/plant assessment tool (5Cs framework)	580–622 kgCO_2/t cement (global benchmarks); case plant overall score 3.501	Structured scoring enables prioritization; plant case shows on-track decarbonization via AF, SCMs, controls	Khalil and AbouZeid [33]
Cement/Provincial China (Henan) co-control bottom-up	Sector 2022: 54.84 Mt CO_2; 2050 IS: 80% CO_2 and LAP reduction versus 2022	Demand/capacity cuts dominate; energy measures deliver LAP benefits; CCUS needed for neutrality but with trade-offs	Zhang et al. [34]

(continued)

Table 1 (continued)

Industry and process	CO_2 intensity	Key findings	References
Cement/solar calciner integration (Chile)	Specific emissions could drop to 630–661 $kgCO_2/t$ clinker by 2050 (reduction of 25 to 21% from 839 $kgCO_2/t$ clinker)	Solar calcination yields notable cuts; strongest with higher carbon price and paired with CCUS	Marin et al. [35]
Petrochemicals/ ethylene oxide via CO_2 electrolysis (tandem: $CO_2 \rightarrow C_2H_4 \rightarrow$ EO)	2.6–10.2 tCO_2e/t EO (cradle-to-gate), depending on electricity (0–0.128 $kgCO_2e/kWh$); conventional EO $\approx$ 2.32 tCO_2e/t EO	Low-carbon power (< 0.04 $kgCO_2e/kWh$) and $\geq$ 50% performance improvement can beat fossil EO; prioritize higher C_2H_4 FE, lower energy demand, higher CO_2 SPCE	Badewole et al. [36]
Petrochemicals/EO via electrocatalytic $C_2H_4 + H_2O_2$ from CO_2 (TEA)	Cradle-to-gate GHG: up to 0.15–0.49 $kgCO_2e/kg$ EO with biogenic CO_2 and RES; higher than fossil if grid-average power	Economic and GHG viability hinges on energy efficiency, low-cost RES, high FLH; byproduct credits and ETS improve outlook	Rodin et al. [37]
Petrochemicals/ ethylene (bioethylene) from U.S. corn ethanol (standalone dehydration and FCC co-processing), cradle-to-gate	Fossil ethylene: 0.9–1.7 $kgCO_2e/kg$ (U.S. avg. $\approx$ 1.1). Bioethylene production 0.4 $kgCO_2e/kg$ (standalone routes); FCC co-processing: 0.3 (C-14 yields) to 0.0 $kgCO_2e/kg$	Bioethylene can be below fossil ethylene due to biogenic C credit; lowest with RNG from manure or stover CHP; linear relation between ethanol CI and ethylene CI	Benavides et al. [38]

challenges from rising demand: global cement production is projected to grow to 20 billion m^3 annually by 2050, while plastics production could triple by 2060, further entrenching emissions trajectories without systemic intervention [27, 40]. The emissions profile of heavy industries bifurcates into energy-related emissions (from fossil fuel combustion) and process emissions (inherent to chemical reactions). In steelmaking, 75% of emissions are energy-related, tied to coal-dependent BF-BOF processes, whereas cement's process emissions dominate at 50–70% [19, 26]. Petrochemicals exhibit a hybrid profile: naphtha cracking emits CO_2 both from energy use (furnace heating) and chemical reactions (feedstock decomposition) [39, 40]. Aluminum production, another energy-intensive sector, emits 1.1 billion tonnes of CO_2 annually, primarily from coal-powered electricity for electrolysis [19]. Regional disparities further shape emissions; Asia's steel and cement sectors, responsible for 70% of global output, rely on young coal-fired infrastructure, whereas Europe and North America show incremental progress via EAFs and alternative fuels [19, 26]. Decarbonizing these sectors necessitates tailored strategies: hydrogen-based direct reduced iron (H_2-DRI) and CCUS for steel; clinker substitution and electrified kilns for cement; and bio-based feedstocks or chemical recycling for plastics [19, 27, 40]. However, achieving the IEA's net-zero scenario requires 60% reductions in steel's emission intensity by 2050 and 90% capture of cement process emissions, underscoring the scale of innovation and investment needed to align heavy industries with climate goals [19, 26].

3 Industry-Specific Challenges in Decarbonization

Decarbonization is a critical goal for industries worldwide, as it aligns with global efforts to mitigate climate change and achieve net-zero emissions. However, different industries face unique challenges due to their operational processes, energy requirements, and technological constraints. Within this landscape, cement and concrete stand out as major contributors responsible for close to one tenth of all energy related carbon dioxide release. The central challenge arises in clinker manufacture, a process that consumes vast amounts of energy and for which broadly scalable substitutes are still lacking [41]. Oil refining in particular is experiencing mounting pressure to cut its carbon footprint as global energy systems gradually shift away from fossil resources. This pressure is expressed through multiple challenges, including the task of curbing emissions generated during refining, advancing the production of cleaner fuels, and finding ways to incorporate renewable power into everyday operations. At the same time, regulatory inconsistencies together with infrastructure gaps create further delays in the movement toward zero emission technologies [42]. Challenges include reducing emissions from farming practices, optimizing energy use in manufacturing, and minimizing waste throughout the supply chain. The industry also faces behavioural barriers, such as consumer preferences and the need for sustainable packaging solutions [43]. The maritime industry is significant contributor to global emissions, with challenges centered on adoption of zero-emission vessel (ZEV) technologies. Major challenges include the heavy financial burden of initial investments, limited infrastructure for alternative fuels, and a regulatory framework that remains fragmented. On the operational side, issues tied to the driving range and overall performance of zero emission vehicles add another layer of complexity to the broader transition toward a low carbon future [44]. Energy-intensive industries, such as steel, mining, and pulp and paper, face common challenges, including the need for electrification, fuel switching, and the deployment of CCS technologies. These industries require long-term policy support, market creation for low-carbon materials, and significant investment in research and development to achieve deep decarbonization [45].

4 Technological Innovations for Decarbonization

4.1 Carbon Capture, Utilization, and Storage (CCUS)

CCUS is frequently viewed as one of the most viable strategies for achieving deep reductions in emissions within sectors where process related carbon dioxide release cannot be entirely avoided, most notably in cement, steel, and chemical production. The technology comprises three key steps: capturing CO_2 emissions at the source, transporting the concentrated CO_2 via pipeline or ship, and then either permanently storing it in geological formations (such as deep saline aquifers or depleted oil and gas reservoirs) or converting it into valuable products. In the cement sector, for instance, roughly 60% of total emissions result from the calcination of limestone, a chemical process that inherently releases CO_2. Retrofitting kilns with capture systems can intercept up to 90% of those emissions, preventing millions of tonnes of CO_2 from entering the atmosphere each year [46]. In steelmaking, blast-furnace off-gases can be captured post-combustion, or pre-combustion capture can be applied in integrated gasification combined cycle (IGCC) plants, reducing emissions by more than half. Captured CO_2 can then be sequestered underground or used as a feedstock in products such as carbon-enhanced concrete, synthetic fuels, or polymers, closing the loop on carbon and creating new revenue streams. Despite its promise, CCUS faces significant hurdles: high capital and operating costs (capture alone can add $40–$100 per tonne of CO_2), limited CO_2 transport and storage infrastructure, and the need for clear regulatory frameworks and liability rules to underwrite long-term storage. Nonetheless, with targeted incentives—such as tax credits, carbon pricing, and public–private partnerships CCUS can play an indispensable role in achieving near-zero emissions in hard-to-abate sectors [47] (Fig. 1).

4.2 Hydrogen Fuel

Hydrogen fuel, especially when produced via electrolysis powered by renewable energy (so-called "green hydrogen"), is another cornerstone technology for decarbonizing energy-intensive industries. In steel production, green hydrogen can replace coke and natural gas as reducing agent in direct reduced iron (DRI) processes, cutting CO_2 emissions by 95% compared to traditional blast-furnace routes. In chemical plants, hydrogen is both an energy carrier and a critical feedstock for ammonia and methanol synthesis; switching from grey hydrogen (derived from steam-methane reforming) to green hydrogen eliminates CO_2 emissions from hydrogen production altogether. Moreover, hydrogen can supply high-temperature process heat (up to 1200 °C) when combusted in turbines or burners, making it suitable for industries such as glass, ceramics, and refining. However, the widespread adoption of hydrogen faces three principal challenges: scaling up electrolyzer capacity to gigawatt levels; reducing the cost of green hydrogen from today's $4–6 per kilogram to below $2/kg;

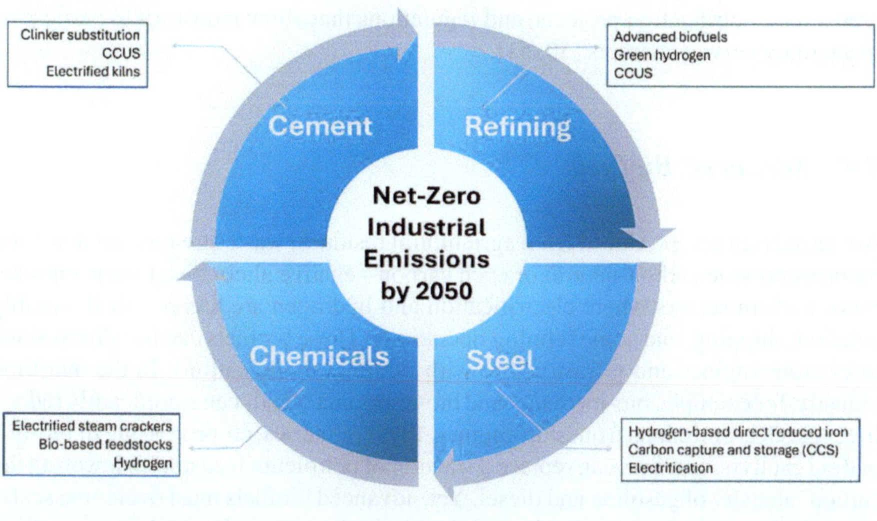

Fig. 1 This figure illustrates the key decarbonization pathways required to achieve net-zero emissions in heavy industries, including steel, cement, chemicals, and refining [48]

and building the infrastructure—pipelines, storage caverns, and refueling stations—needed to transport and distribute hydrogen safely and efficiently [49]. Addressing these challenges will require coordinated policy support in the form of production subsidies, long-term power purchase agreements for renewables, and streamlined permitting for hydrogen networks.

4.3 Electrification and Renewable Energy

Electrifying industrial processes and powering them with renewable electricity is a powerful lever for cutting CO_2 emissions. Electric arc furnaces (EAFs), for example, use electricity to melt steel scrap at temperatures exceeding 1600 °C, reducing CO_2 emissions by roughly 75% relative to coal-fired blast furnaces—provided the electricity comes from low-carbon sources. In the chemical industry, resistive and induction heating can replace fossil-fuel boilers for many medium-temperature processes, while industrial heat pumps can efficiently deliver low- to medium-temperature heat for drying, distillation, and evaporation. Renewable electricity sourced from on-site solar panels, wind turbines, or grid-connected renewables eliminates scope 2 emissions and can be coupled with demand-response strategies to align industrial energy use with periods of high renewable output. Nonetheless, electrification at scale demands substantial investment in grid infrastructure, including high-capacity transmission lines, substation upgrades, and energy-storage systems to manage variability. It also requires supportive policies such as time-of-use electricity tariffs, grants for electrification projects, and regulations that allow industries to participate in ancillary service markets [50, 51].

4.4 Advanced Biofuels

Advanced biofuels produced from agricultural residues, waste streams and non-food biomass provide carbon-neutral or even carbon-negative alternative to conventional fossil fuels in sectors where electrification and hydrogen are less practical, notably aviation, shipping, and some refining operations. These biofuels can be "dropped in" to existing engines and infrastructure with minimal modifications. In the maritime industry, for example, bio-methanol and bio-ammonia blends can significantly reduce lifecycle CO_2 emissions from ship engines. In oil refineries, co-processing of bio-oils in fluid catalytic crackers can replace a fraction of petroleum feedstocks, lowering the carbon intensity of gasoline and diesel. Yet, advanced biofuels must overcome scalability challenges: securing reliable, sustainable feedstock supplies without competing with food production or triggering land-use change; improving conversion efficiencies to reduce costs; and building distribution and blending infrastructure. Lifecycle analyses indicate that well-managed biofuel pathways can achieve up to 80% GHG reductions relative to fossil fuels, but supportive policies such as blending mandates, low-carbon fuel standards, and feedstock sustainability criteria are essential to scale production and ensure environmental [44, 47].

4.5 Circular Economy and Resource Efficiency

Beyond decarbonizing energy supply, reducing material and resource intensity through circular-economy strategies is critical for cutting industrial emissions. Recycling metals, plastics, and other materials can slash energy requirements by 50–90% compared to primary production. In the chemical sector, waste-to-chemicals technologies convert industrial and municipal waste streams into feedstocks, closing material loops and avoiding emissions from both virgin production and landfill methane. In cement, using recycled construction and demolition waste as aggregate or incorporating industrial by-products such as fly ash and slag can reduce clinker demand and thus process emissions by up to 40%. Additionally, optimizing production processes through lean manufacturing, heat integration, and solvent recovery can deliver further energy savings. For circularity to deliver its full potential, production systems need to adopt design for recycling approaches, supported by standardized material flows and responsibility schemes that extend to the producer. Such mechanisms can create strong incentives for manufacturers to reclaim and recycle products once they reach the end of their useful life [23, 50].

4.6 Digitalization and Smart Technologies

Emerging digital tools including artificial intelligence, the Internet of Things, digital twins, and advanced analytics are reshaping industrial operations by making possible real time monitoring, predictive maintenance, and improved process control. AI algorithms can analyze vast datasets from sensors to identify energy-saving opportunities, such as optimizing furnace temperatures or adjusting compressor speeds. Digital twins, which function as virtual counterparts to physical assets, enable engineers to model process adjustments, experiment with control approaches, and anticipate maintenance needs without interrupting production. For example, AI-driven optimization in chemical plants has achieved energy savings of 5–10%, while predictive-maintenance systems have reduced unplanned downtime by up to 30%, indirectly lowering emissions associated with startups and shutdowns [52]. Realizing these benefits requires investment in cybersecurity, interoperable data standards, and workforce training programs that strengthen digital competencies. Collectively, six technological pathways CCUS, hydrogen fuel, electrification with renewables, advanced biofuels, circular economy approaches, and digitalization provide a comprehensive framework for industrial decarbonization. Each technology addresses different facets of the challenge: CCUS tackles unavoidable process emissions, hydrogen and electrification decarbonize high temperature heat and power, biofuels substitute for liquid and gaseous fuels in hard to electrify sectors, circularity reduces material related emissions and digitalization accelerates optimization and innovation. The pace and scale of deployment will depend on coordinated policy frameworks carbon pricing, subsidies, mandates, and infrastructure planning as well as public-private partnerships. By integrating these innovations into a cohesive strategy, industries can achieve deep, sustained cuts in greenhouse gas emissions while maintaining competitiveness and supporting broader economic and environmental goals.

5 Policy and Regulatory Frameworks

Sound policy and regulatory frameworks remain indispensable for both enabling and accelerating pathways of industrial decarbonization. At their foundation, such frameworks establish the economic and legal conditions that guide investment toward low carbon technologies, stimulate innovation, and promote fairness across industrial sectors. One of the most powerful tools in this arsenal is carbon pricing, which internalizes the external cost of greenhouse-gas emissions by assigning a fee to each tonne of CO_2 emitted. Whether implemented as a carbon tax or a cap-and-trade system, carbon pricing sends a clear market signal: technologies and processes that reduce emissions become more economically attractive relative to their high-carbon counterparts. For example, a sufficiently high carbon price can tip the balance in favor of investing in carbon capture, utilization, and storage (CCUS) for cement kilns or green hydrogen for steel production, even when these options carry higher upfront costs. Complementing carbon pricing are subsidies and financial incentives designed to lower the capital and operating costs of emerging clean technologies. Instruments such as grants, tax incentives, and concessional loans can render first-of-its-kind projects viable, helping bridge the difficult gap between pilot demonstration and full commercialization. In shipping, for example, uptake of zero emission vessels has been slowed by substantial capital requirements and uncertain access to alternative fuels [44, 46]. Practical collaboration such as Mission innovation unites countries and private stakeholders to increase funding for clean energy research. These cooperative efforts help reduce duplication of effort, lower costs through shared learnings, and create global markets for low carbon technologies [23, 52]. For instance, a nation with well-established CCUS capabilities could work with cement-producing countries to jointly design capture plants, exchange operational knowledge, and craft regulatory structures suited to their specific geological settings. Such collaborations not only speed up deployment but also ensure that best practices on everything from community engagement to long term monitoring of CO_2 storage sites are disseminated widely. Public private partnerships (PPPs) provide an effective vehicle for mobilizing the considerable finance and technical expertise needed to deliver large scale decarbonization projects [47]. This approach has been successfully applied in renewable energy projects and is increasingly being explored for industrial decarbonization. For example, a blended-finance model could support a CCUS hub serving multiple steel and cement plants, lowering the unit cost of CO_2 transport and storage and making the business case more compelling for all participants.

6 Case Studies and Emerging Examples of Industrial Sustainability Practices

6.1 Direct Reduced Iron in Steel Production

Unlike conventional blast furnace methods which depend on coking coal, Direct Reduced Iron (DRI) process lower carbon emissions by using hydrogen as a reducing agent for iron ore [53, 54]. A case study of an integrated Hydrogen-based DRI–Electric Arc Furnace (H2 DRI-EAF) steel plant demonstrates the real-world feasibility of green steel production, employing a hybrid mix of photovoltaic and wind energy to generate green hydrogen on-site. The hydrogen production system is closely integrated with the steelmaking process, forming a closed-loop model of sustainability. And deployment of power-to-hydrogen conversion systems alongside carbon capture and storage (CCS) technologies enhances the environmental performance of the facility, achieving near-zero emissions while maintaining industrial-scale productivity [55, 56]. Moving from traditional steelmaking toward direct reduced iron processes driven by green hydrogen requires very large initial investments. These costs are primarily associated with building new infrastructure for hydrogen production, acquiring advanced electrolyzers, and implementing carbon capture and storage facilities. However, despite the high capital expenditures the economic case for adopting these technologies is gaining traction. In the medium to long term the operational cost savings derived from improved energy efficiency, lower carbon compliance costs, and reduced environmental liabilities make green steel a financially viable alternative [55, 57]. Moreover, the emergence of a green steel market is expected to stimulate economic growth across sectors aligned with sustainability, encompassing the development of hydrogen supply chains, renewable energy production, and environmental engineering services. By eliminating the use of coking coal and replacing it with hydrogen sourced from renewable energy, direct CO_2 emissions from steelmaking can be reduced by as much as 98% compared to conventional blast furnace methods [58]. This dramatic reduction addresses global climate targets and significantly improves local air quality by limiting emissions of sulfur oxides, nitrogen oxides, and particulate matter. Integrating CCS further enhances sustainability by capturing residual emissions, allowing steel plants to approach net-zero operation without compromising productivity or economic output.

From technological feasibility demonstrated in real-world case studies to strong policy backing from entities like the European Union, the momentum toward green steel production is building rapidly. Although the initial economic investment is significant, the long-term benefits economic, environmental, and social make this a worthwhile and strategic direction for future growth. With expected improvements in hydrogen production efficiency and further innovations in CCS technology, the cost-effectiveness of green steel is likely to improve over time, making it more accessible to a wider range of producers. The combined effect of technological advancements, supportive policies, and increasing market demand places DRI with green hydrogen at the forefront of sustainable industrial transformation.

6.2 Innovations in Low-Clinker Cement and Carbon Capture Technologies

The cement industry, a cornerstone of global infrastructure and development, is also one of largest industrial emitters of carbon dioxide, accounting for about 8% of global CO_2 emissions [59]. In response to the growing pressure to decarbonize, significant technological innovations are being developed and implemented to reduce the carbon footprint of cement production. Two key approaches gaining traction are the development of low-clinker formulations and the integration of carbon capture and storage (CCS) technologies. Traditional cement manufacturing processes are highly carbon-intensive due to the calcination of limestone to produce clinker the key binding component of cement. This process not only consumes large amounts of energy, usually derived from fossil fuels, but also releases CO_2 as a byproduct of the chemical reaction itself. To counter this challenge, the adoption of supplementary cementitious materials including blast furnace slag, silica fume and fly ash has gained traction as a promising avenue. These materials partially replace clinker in cement formulations, thereby lowering both the thermal energy requirement and the CO_2 emissions associated with production [41, 60]. A Norwegian cement facility demonstrates how decarbonization can be achieved by reducing clinker content by 30%, which lowered CO_2 emissions by 20% [60]. Policymakers play a central role in enabling such transformations particularly in hard to abate sectors. Carbon pricing is among the most effective tools, as seen in the European Union Emissions Trading System, where placing a financial value on carbon has driven industries to cut emissions while managing compliance costs [41]. Although the adoption of CCS and low-clinker formulations requires significant capital investment, the long-term outlook is favorable. Lower emissions reduce exposure to carbon pricing, while demand for sustainable construction materials continues to expand under regulatory and market pressures. Since clinker is the main source of emissions, reducing its share directly cuts CO_2 output, while SCMs improve both performance and durability. When coupled with CCS, plants can reduce emissions by up to 50% compared with conventional practices [41, 60]. These integrated strategies further lower energy use and raw material extraction by incorporating industrial by-products such as fly ash and slag, while CCS prevents additional atmospheric release, positioning the cement industry to contribute meaningfully to global climate targets.

6.3 Advancing Sustainability in Plastics: Chemical Recycling and Bio-based Feedstocks

The plastics sector, long reliant on fossil feedstocks, is moving toward circularity through chemical recycling and bio-based feedstocks [61, 62]. Unlike mechanical recycling, which downgrades quality, chemical recycling recovers virgin-grade plastics and is particularly valuable for mixed or contaminated waste streams. This reduces both plastic pollution and dependence on fossil inputs [62]. Bio-based plastics, derived from biomass carbon, also cut lifecycle emissions, and although not all are biodegradable, many are designed to degrade under composting or natural conditions [61]. BASF's advanced recycling facility illustrates this shift, using depolymerization to convert mixed waste into monomers that are re-polymerized into high-quality plastics, thereby lowering fossil demand and reducing waste accumulation [62]. The diffusion of such innovations is closely tied to regulation. Extended Producer Responsibility (EPR) policies require manufacturers to manage full lifecycle of plastics, while recycled content mandates, standards for compostable plastics, and restrictions on single-use products reinforce adoption [61]. These measures create incentives and provide the infrastructure needed for large-scale recycling and bio-based alternatives. While capital costs remain high, the economic rationale strengthens as companies benefit from lower raw material expenses, reduced waste management costs, and compliance with tightening regulations. Firms adopting sustainable plastics can also capture premium markets as consumer demand rises, supported by increasing financial flows into green industries [61, 62]. Chemical recycling diverts large volumes of waste from landfills and oceans while conserving resources, and bio-based plastics can reduce lifecycle CO_2 emissions by up to 80%, owing to biomass carbon sequestration and lower energy intensity [61, 62]. In addition, many bio-based plastics offer favorable end-of-life profiles through biodegradability reducing persistence in natural ecosystems. Despite trade-offs such as land use concerns, their environmental profile remains significantly better than petrochemical alternatives.

6.4 Industrial Clusters Implementing CCUS and Circularity

Industrial clusters are becoming pivotal in decarbonization as networks of energy-intensive industries integrate Carbon Capture, Utilization, and Storage (CCUS) with circular practices to increase efficiency [63, 64]. CCUS enables emissions from refineries, chemical plants, and power stations to be captured and either converted into products such as synthetic fuels and construction materials or permanently stored, while circular strategies extend material lifecycles through reuse and recycling [63]. The Rotterdam cluster exemplifies this integrated model by capturing emissions from refineries and chemical plants, converting part into carbon-neutral fuels and storing the remainder offshore, while promoting by-product exchanges and shared infrastructure to strengthen material efficiency [63, 64]. Recognizing the scale of investment required, governments are easing risks through incentives and programs such as the EU's Horizon Europe initiative, which funds CCUS and circular economy innovations [64, 65]. Carbon pricing schemes and performance standards further ensure that industries face strong financial and strategic drivers to adopt these approaches. The environmental benefits are substantial, with Rotterdam achieving emission reductions of up to 50% through combined CCUS and circularity measures [63, 64].

7 Conclusion

Decarbonizing heavy industries steel, cement, chemicals, and refining demands urgent action to overcome entrenched reliance on fossil fuels, process emissions, and long asset lifespans. Key challenges include high costs of emerging technologies like hydrogen-based steelmaking and carbon capture, infrastructure gaps for renewable energy and CO_2 transport, and policy inertia. However, opportunities abound: hydrogen-DRI, electrified processes, and circular economy strategies (e.g., material recycling, industrial symbiosis) can drastically reduce emissions while enhancing efficiency. Near-zero pathways require rapid scaling of proven solutions (e.g., CCUS-equipped cement plants), policy alignment (carbon pricing, green procurement), and digital tools (AI-driven optimization, supply chain tracking). Stakeholders must act decisively: governments should enforce carbon border adjustments and fund R&D;

industries must adopt science-based targets and low-carbon procurement; investors must prioritize green infrastructure; academia should advance next-gen technologies; and civil society must advocate for equitable transitions. Global collaboration through initiatives like Mission Innovation and the Paris Agreement's NDCs is critical to harmonize standards, share knowledge, and accelerate deployment. The window to limit warming to 1.5 °C is narrowing, but the convergence of technological innovation, policy ambition, and cross-sector partnerships offers a viable roadmap to redefine industrial progress within planetary boundaries. Success hinges on collective urgency, innovation, and inclusivity to ensure a just transition for workers and communities reliant on fossil-dependent industries.

References

1. Kharissova AB, Kharissova OV, Kharisov BI, Méndez YP (2024) Carbon negative footprint materials: a review. Nano-Struct Nano-Objects 37:101100. https://doi.org/10.1016/j.nanoso.2024.101100
2. Rogelj J, den Elzen M, Höhne N, Fransen T, Fekete H, Winkler H, Schaeffer R, Sha F, Riahi K, Meinshausen M (2016) Paris agreement climate proposals need a boost to keep warming well below 2 °C. Nature 534(7609):631–639. https://doi.org/10.1038/nature18307
3. Bataille C, Waisman H, Briand Y, Svensson J, Vogt-Schilb A, Jaramillo M, Delgado R, Arguello R, Clarke L, Wild T, Lallana F, Bravo G, Nadal G, Le Treut G, Godinez G, Quiros-Tortos J, Pereira E, Howells M, Buira D et al (2020) Net-zero deep decarbonization pathways in Latin America: challenges and opportunities. Energy Strat Rev 30:100510. https://doi.org/10.1016/j.esr.2020.100510
4. OECD (2019) Global Material resources outlook to 2060: economic drivers and environmental consequences. https://doi.org/10.1787/9789264307452-en
5. Holappa L (2020) A general vision for reduction of energy consumption and CO_2 emissions from the steel industry. Metals 10(9):1–21. https://doi.org/10.3390/met10091117
6. Pisciotta M, Pilorgé H, Feldmann J, Jacobson R, Davids J, Swett S, Sasso Z, Wilcox J (2022) Current state of industrial heating and opportunities for decarbonization. Prog Energy Combust Sci 91:100982. https://doi.org/10.1016/j.pecs.2021.100982
7. Kim J, Sovacool BK, Bazilian M, Griffiths S, Lee J, Yang M, Lee J (2022) Decarbonizing the iron and steel industry: a systematic review of sociotechnical systems, technological innovations, and policy options. Energy Res Soc Sci 89:102565. https://doi.org/10.1016/j.erss.2022.102565
8. Kabeyi MJB, Olanrewaju OA (2022) Sustainable energy transition for renewable and low carbon grid electricity generation and supply. Front Energy Res 9. https://doi.org/10.3389/fenrg.2021.743114
9. Kumar A, Tiwari AK, Milani D (2024) Decarbonizing hard-to-abate heavy industries: current status and pathways towards net-zero future. Process Saf Environ Prot 187:408–430. https://doi.org/10.1016/j.psep.2024.04.107
10. Nilsson A, Singh N (2023) The 5 shifts needed to decarbonize industry. World Resources Institute. https://www.wri.org/insights/shifts-to-decarbonize-industry
11. Leicher J, Giese A, Wieland C (2024) Electrification or hydrogen? The challenge of decarbonizing industrial (high-temperature) process heat. Multidiscip Sci J 7(4):439–456. https://doi.org/10.3390/j7040026
12. Marocco P, Gandiglio M, Audisio D, Santarelli M (2023) Assessment of the role of hydrogen to produce high-temperature heat in the steel industry. J Clean Prod 388:135969. https://doi.org/10.1016/j.jclepro.2023.135969

13. Tasleem S, Bongu CS, Krishnan MR, Alsharaeh EH (2024) Navigating the hydrogen prospect: a comprehensive review of sustainable source-based production technologies, transport solutions, advanced storage mechanisms, and CCUS integration. J Energy Chem 97:166–215. https://doi.org/10.1016/j.jechem.2024.05.022
14. McLaughlin H, Littlefield AA, Menefee M, Kinzer A, Hull T, Sovacool BK, Bazilian MD, Kim J, Griffiths S (2023) Carbon capture utilization and storage in review: sociotechnical implications for a carbon reliant world. Renew Sustain Energy Rev 177:113215. https://doi.org/10.1016/j.rser.2023.113215
15. Krauklis AE, Karl CW, Gagani AI, Jørgensen JK (2021) Composite material recycling technology—state-of-the-art and sustainable development for the 2020s. J Compos Sci 5(1). https://doi.org/10.3390/jcs5010028
16. IEA (2024) CO_2 emissions in 2023, vol 24. International Energy Agency. https://www.iea.org/reports/co2-emissions-in-2023
17. Ritchie H (2024) Sector by sector: where do global greenhouse gas emissions come from? Our World in Data. https://ourworldindata.org/ghg-emissions-by-sector
18. Hoffmann C, Van Hoey M, Zeumer B (2020) Decarbonization challenge for steel. https://www.mckinsey.com/industries/metals-and-mining/our-insights/decarbonization-challenge-for-steel#/
19. IEA (2020) Iron and steel technology roadmap. https://www.iea.org/reports/iron-and-steel-technology-roadmap
20. Andrade C, Desport L, Selosse S (2024) Net-negative emission opportunities for the iron and steel industry on a global scale. Appl Energy 358:122566. https://doi.org/10.1016/j.apenergy.2023.122566
21. Rippy K, Bell RT, Leick N (2024) Chemical and electrochemical pathways to low-carbon iron and steel. Npj Mater Sustain 2(1):33. https://doi.org/10.1038/s44296-024-00036-6
22. Gao J, Wang B, Teng F, Qi Y, Zhang Y (2024) Analysis of carbon emission reduction capacity of hydrogen-rich oxygen blast furnace based on renewable energy hydrogen production. High Temp Mater Process 43(1):20240050. https://doi.org/10.1515/htmp-2024-0050
23. Daehn K, Basuhi R, Gregory J, Berlinger M, Somjit V, Olivetti EA (2022) Innovations to decarbonize materials industries. Nat Rev Mater 7(4):275–294. https://doi.org/10.1038/s41578-021-00376-y
24. Subran L, Zimmer M (2023) Greendustry. In: Investing in a changing climate: navigating challenges and opportunities. Springer Nature Switzerland, pp 81–116. https://doi.org/10.1007/978-3-031-47172-8_5
25. Fritzeen WE, O'Rourke PR, Fuhrman JG, Colosi LM, Yu S, Shobe WM, Doney SC, McJeon HC, Clarens AF (2023) Integrated assessment of the leading paths to mitigate CO_2 emissions from the organic chemical and plastics industry. Environ Sci Technol 57(49):20571–20582. https://doi.org/10.1021/acs.est.3c05202
26. IEA (2018) Technology roadmap—low-carbon transition in the cement industry—analysis. IEA. https://www.iea.org/reports/technology-roadmap-low-carbon-transition-in-the-cement-industry
27. Purton M (2024) 4 ways to make the cement industry more sustainable. Sustainable Development. https://www.weforum.org/stories/2024/09/cement-production-sustainable-concrete-co2-emissions/
28. Xia B, Liu Z, Peng J (2025) Experimental study on the effect of hot steam on the physicochemical properties and mineralization efficiency of blast furnace slag–CO_2. Environ Technol Innov 40(2025):1–19. https://doi.org/10.1016/j.eti.2025.104426
29. Chang Y, Wan F, Yao X, Wang J, Han Y, Li H (2023) Influence of hydrogen production on the CO_2 emissions reduction of hydrogen metallurgy transformation in iron and steel industry. Energy Rep 9(2023):3057–3071. https://doi.org/10.1016/j.egyr.2023.01.083
30. Perpiñán J, Bailera M, Peña B, Kannan P, Eveloy V, Romeo LM (2023) Power to gas and top gas recycling integration in an oxygen blast furnace steelmaking industry. J CO_2 Util 78(2023):1–10. https://doi.org/10.1016/j.jcou.2023.102634

31. Massoumi Nejad B, Enferadi S, Andrew R (2025) A comprehensive analysis of process-related CO_2 emissions from Iran's cement industry. Clean Environ Syst 16(2025):1–11. https://doi.org/10.1016/j.cesys.2024.100251
32. He F, Ma J, Hu Q, Wang J, Chen Y, Yang H, Yang Y (2025) Enhancing energy efficiency and decarbonization of cement production through integrated calcium-looping and methane dry reforming (CaL-DRM) for in-situ CO_2 conversion to syngas. Carbon Capture Sci Technol 14(2025):1–13. https://doi.org/10.1016/j.ccst.2024.100359
33. Khalil E, AbouZeid M (2025) A global assessment tool for cement plants improvement measures for the reduction of CO_2 emissions. Results Eng 26(2025):1–19. https://doi.org/10.1016/j.rineng.2025.104767
34. Zhang Y, Wang L, Zhang J (2025) Evaluating the co-control potential of CO_2 and air pollutant emission reductions in the cement industry in China: a case in Henan. J Clean Prod 518(2025):1–13. https://doi.org/10.1016/j.jclepro.2025.145912
35. Marin JSZ, Bern G, Platzer W, Dinter F (2025) Techno-economic analysis of a solar calciner for CO_2 emissions reduction in the Chilean cement industry. Sol Energy 288(2–25):1–16. https://doi.org/10.1016/j.solener.2025.113266
36. Badewole A, Huang JE, Sargent EH, Saville BA, MacLean HL (2025) Carbon footprint assessment of ethylene oxide production via CO_2 electrolysis. RSC Sustain 3(7):3095–3108. https://doi.org/10.1039/D5SU00258C
37. Rodin V, Böhm H, Lindorfer J, Paulik C (2023) Techno-economic evaluation of the electrochemical production of renewable ethylene oxide from fluctuating power sources and CO_2. J CO_2 Util 75(2023):1–20. https://doi.org/10.1016/j.jcou.2023.102554
38. Benavides PT, Gracida-Alvarez UR, Richa K, Port J, Hawkins TR (2025) Cradle-to-gate greenhouse gas emissions of the production of ethylene from U.S. corn ethanol and comparison to fossil-derived ethylene production. Bioresour Technol 430(2025):1–11. https://doi.org/10.1016/j.biortech.2025.132565
39. Bower G, Gaffney M, Adeyemo M, King B (2025) Assessing the greenhouse gas emissions of the US petrochemical industry. https://rhg.com/research/petrochemicals-emissions/
40. Furusawa Y (2024) Pathway to decarbonization of petrochemicals in Japan. Renewable Energy Institute, pp 1–43
41. Griffiths S, Sovacool BK, Furszyfer Del Rio DD, Foley AM, Bazilian MD, Kim J, Uratani JM (2023) Decarbonizing the cement and concrete industry: a systematic review of socio-technical systems, technological innovations, and policy options. Renew Sustain Energy Rev 180:1–10. https://doi.org/10.1016/j.rser.2023.113291
42. Griffiths S, Sovacool BK, Kim J, Bazilian M, Uratani JM (2022) Decarbonizing the oil refining industry: a systematic review of sociotechnical systems, technological innovations, and policy options. Energy Res Soc Sci 89:102542. https://doi.org/10.1016/j.erss.2022.102542
43. Sovacool BK, Bazilian M, Griffiths S, Kim J, Foley A, Rooney D (2021) Decarbonizing the food and beverages industry: a critical and systematic review of developments, sociotechnical systems and policy options. Renew Sustain Energy Rev 143:110856. https://doi.org/10.1016/j.rser.2021.110856
44. Ibokette AI, Ogundare TO, Akindele JS, Anyebe AP, Okeke RO (2024) Decarbonization strategies in the U.S. Maritime industry with a focus on overcoming regulatory and operational challenges in implementing zero-emission vessel technologies. Int J Innov Sci Res Technol 9(11):131–162. https://doi.org/10.38124/ijisrt/IJISRT24NOV829
45. Nurdiawati A, Urban F (2021) Towards deep decarbonisation of energy-intensive industries: a review of current status, technologies and policies. Energies 14(9). https://doi.org/10.3390/en14092408
46. Buznitski K, Verma S, Nitzsche M (2024) Decarbonizing industry: policy approaches to eliminate hard-to-abate emissions. MIT Sci Policy Rev 5(1):58–70. https://doi.org/10.38105/spr.nbbx7l69fu
47. Nurdiawati A, Urban F (2022) Decarbonising the refinery sector: a socio-technical analysis of advanced biofuels, green hydrogen and carbon capture and storage developments in Sweden. Energy Res Soc Sci 84:102358. https://doi.org/10.1016/j.erss.2021.102358

48. International Energy Agency (2021) Net zero by 2050: a roadmap for the global energy sector. https://www.iea.org/reports/net-zero-by-2050
49. Griffiths S, Sovacool BK, Kim J, Bazilian M, Uratani JM (2021) Industrial decarbonization via hydrogen: a critical and systematic review of developments, socio-technical systems and policy options. Energy Res Soc Sci 80:102208. https://doi.org/10.1016/j.erss.2021.102208
50. Kim J, Sovacool BK, Bazilian M, Griffiths S, Yang M (2024) Energy, material, and resource efficiency for industrial decarbonization: a systematic review of sociotechnical systems, technological innovations, and policy options. Energy Res Soc Sci 112:103521. https://doi.org/10.1016/j.erss.2024.103521
51. Mallapragada DS, Dvorkin Y, Modestino MA, Esposito DV, Smith WA, Hodge BM, Harold MP, Donnelly VM, Nuz A, Bloomquist C, Baker K, Grabow LC, Yan Y, Rajput NN, Hartman RL, Biddinger EJ, Aydil ES, Taylor AD (2022) Decarbonization of the chemical industry through electrification: barriers and opportunities. ChemRxiv 1(1):1–18. https://doi.org/10.1016/j.joule.2022.12.008
52. Adebayo YA, Ikevuje AH, Kwakye JM, Esiri AE (2024) Energy transition in the oil and gas sector: business models for a sustainable future. Int J Manag Entrep Res 6(10):1–29. https://doi.org/10.51594/ijmer.v6i10.1611
53. Rechberger K, Spanlang A, Sasiain Conde A, Wolfmeir H, Harris C (2020) Green hydrogen-based direct reduction for low-carbon steelmaking. Steel Res Int 91(11):1–12. https://doi.org/10.1002/srin.202000110
54. Roos G (2024) An overview of the arguments in literature, for and against the use of green hydrogen for the production of direct reduced iron. Preprints 1:1–16. https://doi.org/10.20944/preprints202405.1582.v1
55. Benavides K, Gurgel A, Morris J, Mignone B, Chapman B, Kheshgi H, Herzog H, Paltsev S (2024) Mitigating emissions in the global steel industry: representing CCS and hydrogen technologies in integrated assessment modeling. Int J Greenh Gas Control 131:1–14. https://doi.org/10.1016/j.ijggc.2023.103963
56. Zou B, Zhang Y, Chen Q, Hu Q, Hu X, Shi J, Li Z, Wang Q (2024) Low-carbon economic schedule of the H2DRI-EAF steel plant integrated with a power-to-hydrogen system driven by blue hydrogen and green hydrogen. IET Renew Power Gener 18(16):3839–3854. https://doi.org/10.1049/rpg2.13064
57. Weckenborg C, Graupner Y, Spengler TS (2024) Prospective assessment of transformation pathways toward low-carbon steelmaking: evaluating economic and climate impacts in Germany. Resour Conserv Recycl 203:107434. https://doi.org/10.1016/j.resconrec.2024.107434
58. Gervásio H, da Silva LS (2023) Decarbonisation of steel production by the low-carbon DRI-EAF route. Ce/Papers 6(3–4):509–513. https://doi.org/10.1002/cepa.2687
59. Nehdi ML, Marani A, Zhang L (2024) Is net-zero feasible: systematic review of cement and concrete decarbonization technologies. Renew Sustain Energy Rev 191:1–14. https://doi.org/10.1016/j.rser.2023.114169
60. Barbhuiya S, Kanavaris F, Das BB, Idrees M (2024) Decarbonising cement and concrete production: strategies, challenges and pathways for sustainable development. J Build Eng 86:1–28. https://doi.org/10.1016/j.jobe.2024.108861
61. Bataille C (2020) Low and zero emissions in the steel and cement industries barriers, technologies and policies. In: OECD green growth papers, vol 2020/0, no 66. www.oecd.org/greengrowth
62. Chung C, Kim J, Sovacool BK, Griffiths S, Bazilian M, Yang M (2023) Decarbonizing the chemical industry: a systematic review of sociotechnical systems, technological innovations, and policy options. Energy Res Soc Sci 96:102955. https://doi.org/10.1016/j.erss.2023.102955
63. Carmona-Martínez AA, Rontogianni A, Zeneli M, Grammelis P, Birgi O, Janssen R, Di Costanzo B, Vis M, Davidis B, Reumerman P, Rueda A, Jarauta-Córdoba C (2024) Charting the course: navigating decarbonisation pathways in Greece, Germany, The Netherlands, and Spain's industrial sectors. Sustainability 16(14):1–26. https://doi.org/10.3390/su16146176
64. Rattle I, Taylor PG (2023) Factors driving the decarbonisation of industrial clusters: a rapid evidence assessment of international experience. Energy Res Soc Sci 105:1–17. https://doi.org/10.1016/j.erss.2023.103265

65. Szemalikowska A (2024) Advancements in transition to green steel in Europe through direct reduced iron production. Mater Technol 68(2):1–9. https://doi.org/10.32730/mswt.2024.68.2.2

From Agroforestry to Pharmaceuticals: Cross-Sectoral Ethnobotanical Approaches for Carbon-Neutral Industries

Niharika Sharma and Radha

Abstract Ethnobotany plays a significant role in advancing carbon neutrality by combining traditional ecological knowledge with sustainable industrial practices. Indigenous communities, especially in the Western Himalaya, have used agroforestry, rotational farming, and sacred grove preservation to support carbon sequestration, conserve biodiversity, and manage natural resources sustainably. Their climate-resilient crops and wild plant-based food systems also strengthen food security and environmental resilience. Replacing high-emission industrial goods with ethnobotanical alternatives—such as using plant-based dyes instead of synthetic ones, natural fibers for textiles, and biodegradable resins for adhesives—helps lower carbon footprints. Additionally, ethnoveterinary practices like using herbal feed additives can reduce methane emissions and encourage sustainable livestock management. Industries can also use plants to capture carbon by using medicinal and aromatic plants to purify the air, cleaning up pollution with special plants (phytoremediation), and adopting eco-friendly methods to extract plant-based materials while protecting the environment. Traditional healing systems, which include Ayurveda and Traditional Chinese Medicine (TCM), demonstrate sustainable medicinal plant use through regenerative agriculture, ethical harvesting, and eco-friendly processing techniques. To maximize the potential of ethnobotany in carbon neutrality, strong policy agendas and collaborations between policymakers, industries, and indigenous communities are essential. By integrating ethnobotanical principles into industrial strategies, industries can drive low-carbon development, circular economies, and sustainable innovations. The relationship between traditional knowledge and modern sustainability plays a significant role in developing carbon-neutral industries while at the same time preserving cultural heritage and biodiversity.

Keywords Ethnobotany · Carbon neutrality · Sustainable industries · Indigenous knowledge

N. Sharma · Radha (✉)
School of Biological and Environmental Sciences, Shoolini University of Biotechnology and Management Sciences, Solan, India
e-mail: radhuchauhan7002@gmail.com

S. S. Muthu (ed.), *Reduction of Industrial Carbon Footprint*, Environmental Footprints and Eco-design of Products and Processes,
https://doi.org/10.1007/978-3-032-10913-2_3

1 Introduction

For centuries, people have depended on plants for many purposes, including food, medicine, shelter, decoration, construction, and clothing. Indigenous and local communities have vast knowledge of using these plants for various purposes, and using these plants with local traditional knowledge is known as traditional ethnobotanical knowledge. Nowadays, this knowledge is not well documented as the younger generation is not very interested in this traditional knowledge, but has been orally transmitted from one generation to another, and by doing this, the knowledge will one day disappear from the face of the earth [23]. However, to protect this knowledge, ethnobotanists have played an essential role in unravelling and documenting these plant-people interactions and have unlocked the knowledge by conducting various interviews and surveys. While there is not much information and records found about how humans used these plants in ancient times. Whereas, it was recently found that people in South Africa used the leaves of a plant called *Cryptocarya woodii* to make bedding about 77,000 years ago. Scientists were able to figure this out using advanced tools that study ancient plant DNA, called archaeogenetics [48]. This plant is still used by the people who live in the same area where it was discovered, and it was found that it helps keep mosquitoes away because it can kill their larvae. The recovery of starch granules from the surfaces of at least 105,000-year-old stone tools from Mozambique suggests that early humans consumed grass seeds [25]. Ethnobotany is a field that works at the intersection of plants and people at one end and science at the other end; therefore, ethnobotanists can act as a bridge between them [4]. The urgent challenge of climate change has elevated the search of carbon neutrality as a top priority for industries worldwide, requiring inventive and sustainable approaches to control greenhouse gas emissions and enhance the removal of carbon dioxide from the atmosphere [1, 49]. In response, there is a growing appreciation for the integration of ethnobotanical knowledge rooted in traditional plant use and management into contemporary industrial strategies for carbon neutrality. Ethnobotanical approaches provide distinctive avenues toward achieving carbon neutrality by utilizing the carbon-sequestering abilities of plants, encouraging the sustainable use of biomass, and aiding the restoration and preservation of natural ecosystems [7, 49]. Initiatives such as afforestation, producing bioenergy with carbon capture, and incorporating plant-based materials in manufacturing not only decrease dependence on fossil fuels but also promote circular resource flows in industrial systems [1]. The principle of carbon neutrality in biomass systems relies on the natural balance between carbon absorbed by plants during growth and the carbon released when biomass is utilized, emphasizing the need for sustainable management and short harvest cycles to maximize environmental benefits [7]. Utilizing both traditional ecological knowledge and modern technological advancements, ethnobotanical methods can enhance established approaches like renewable energy implementation, carbon capture and storage, and the transformation of waste into valuable resources [1, 49]. Safeguarding and incorporating ethnobotanical knowledge are crucial not only for preserving cultural traditions but also for development innovative, sustainable strategies to address contemporary environmental issues. Integrating traditional knowledge with scientific growth allows industries to adopt more adaptive and climate-conscious practices. Recognizing and revitalizing this ancient knowledge could be contributory in achieving a sustainable and carbon-neutral future.

2 Indigenous Knowledge and Carbon Sequestration

Over the past thirty years, agroforestry has achieved worldwide recognition as a comprehensive approach to sustainable land management, appreciated for its productivity and beneficial effects on the environment. Its recent addition as a greenhouse gas mitigation measure under the Kyoto Protocol has brought greater attention to its role in biological carbon (C) sequestration. This potential arises from the concept that integrated systems like agroforestry make more efficient use of resources—such as nutrients, sunlight, and water—compared to monocultures, resulting in increased overall carbon capture and storage [28]. Capturing carbon from the atmosphere and storing it within terrestrial ecosystems is a recognized strategy for offsetting greenhouse gas emissions. Farmlands, especially when trees are incorporated and managed together with crops and livestock, offer considerable potential for carbon storage. Consequently, agroforestry is gaining acknowledgment not only for advancing sustainable agricultural practices but also for its significant contribution to addressing climate change challenges [2]. Agroforestry involves integrating trees into agricultural landscapes. This practice has proven beneficial for farmers in both developed and developing countries, contributing to enhanced soil health. While agroforestry provides distinct advantages, such as improved productivity and resilience, planting trees in degraded forest areas has also demonstrated effectiveness in fighting climate change by capturing and storing carbon [27]. India hosts thousands of sacred groves-forests preserved by local communities due to their cultural or spiritual importance. These sacred sites are often rich in biodiversity and play a crucial ecological role. Unlike government-managed protected areas, most sacred groves in India are overseen by community groups and are typically situated on private or communal land. This unique form of stewardship presents distinct challenges for future management and policy development concerning the conservation of these ecologically and culturally significant areas [30]. The rising population pressure, along with forest degradation and depletion, makes sustainable forest management a major challenge. In India, forests cover only about 23.8% of the country's total geographical area, placing it among the nations with low forest cover [21]. Many indigenous communities living in and near forests have long practiced the conservation of natural resources and biodiversity as part of their traditions. The idea of sacred groves dates back to ancient times, even mentioned in the Rig-Veda, where tree worship was widespread and deeply rooted in cultural practices. Particularly, the Buddha is said to have attained enlightenment while meditating under a banyan tree, further highlighting the historical significance of trees in spiritual and cultural life [29]. Traditional knowledge has been used for centuries by indigenous local communities to manage natural resources under local laws, customs, and traditions [12]. Across the globe, numerous traditional cultures hold nature in deep reverence, often regarding specific plants and animals as sacred. In the Western Ghats, sacred groves represent small remnants of ancient forests that are set aside in honour of local animistic deities. These groves serve as both spiritual sites and important reservoirs of biodiversity [47]. However, shifts in socio-economic factors and evolving land use practices have put many sacred groves at risk. These changes have led to reductions in their size, as well as alterations in their vegetation structure and the variety of species they support [10].

3 Ethnobotanical Contributions to Sustainable Industry Practices

Ethnobotany bridges scientific and traditional knowledge, enabling the identification of culturally significant and ecologically important plant species. This guides conservation priorities and supports community-driven conservation initiatives, such as the protection of sacred groves and the use of sustainable harvesting techniques [16]. In a study, researchers developed a high-protein (18% w/w) plant-based cheese alternative using waxy starch, plant-protein isolate, and coconut oil. This enhanced viscoelastic properties and meltability, outperforming commercial alternatives in texture and rheological properties [15]. Research indicates that plant-based meat analogues produce a substantially lower environmental footprint compared to beef burgers, resulting in a 62% reduction in greenhouse gas emissions [43, 44]. Life cycle assessments reveal that plant-based patties consistently outperform beef burgers in reducing not only GHG emissions but also water consumption and land use [43, 44]. Similarly, plant-based dairy alternatives have been shown to be both healthy and sustainable, offering reductions of 59–71% in greenhouse gas emissions, as well as lower land and water use relative to dairy milk, all while maintaining nutritional adequacy [13]. Ethnoveterinary practices present valuable approaches for reducing methane emissions in livestock, drawing on traditional knowledge and the use of natural remedies. Rooted in generations of experience, these methods can work alongside modern strategies to address methane emissions from ruminants, which are major sources of greenhouse gases. Research examining various ruminant management methods for enteric methane (CH_4) reduction in both extensive and intensive systems shows that selecting and breeding animals with low residual feed intake can significantly decrease CH_4 emissions. Higher productivity in dairy cows is linked with lower CH_4 emissions per kilogram of milk produced. Additionally, the review emphasizes the importance of diet composition; specifically, increasing concentrate intake and providing low-fiber, high-protein feeds are effective in lowering methane output. The choice of grain source and the proportion of concentrate in the diet are also critical. Overall, digestive physiology, diet quality, and selective breeding emerge as key strategies for methane mitigation in ruminant livestock [8]. Detailed insights into these strategies are presented in Table 1. Further, studies emphasize targeting ruminal fermentation, a key source of methane emissions, due to its large impact on global warming. A spectrum of approaches has been explored: vaccination, enzyme inhibitors, phage therapy, feed processing techniques, supplementation with fatty acids and organic acids, ionophores, microbial additives, non-ionic surfactants, halogenated methane analogs, sulphates, and defaunation, as well as dietary adjustments. Among these, plant secondary metabolites (PSMs) stand out as promising natural options, modifying rumen fermentation and reducing methane output. Managing methane from manure is also recognized as an effective supplementary strategy. Overall, reducing methane emissions supports climate goals while improving feed efficiency and livestock productivity, confirming the central role of rumen microbial management in animal nutrition [3]. Research has shown that methane emissions from ruminants are shaped by a range of factors, including the type of feed, rumen microbial communities, animal genotype, and the quality of the diet. Successful mitigation strategies incorporate nutritional adjustments, careful forage selection, and pasture management methods such as monitoring forage mass and adjusting stocking rates to optimize animal intake and decrease emissions. Important factors of methane production include dry matter intake, dietary composition, and interactions between microbes and the rumen environment. Notably, animals with higher feed efficiency do not always produce less methane, highlighting the importance of understanding the methanogenic characteristics of different genotypes. The study emphasizes the necessity for integrated, system-based research that brings together dietary, production, and management practices to craft effective and durable solutions for methane reduction [18].

4 Ethnobotany and Industrial Carbon Capture

Ethnobotanical research has revealed plant species with notable capacities for carbon sequestration, insights that frequently stem from traditional ecological knowledge. Indigenous and local communities manage landscapes in ways that promote biomass productivity and increased carbon storage—using agroforestry, mixed cropping systems, and conserving high-biomass plants. These time-tested practices can serve as models for large-scale industrial afforestation, reforestation, and land management initiatives focused on maximizing carbon sequestration [46]. In addition, studies assessing CO_2 sequestration technologies and their practical uses in industry highlight that post-combustion and oxy-fuel combustion processes are especially effective for carbon capture in the cement sector, with oxy-fuel combustion achieving higher efficiency. Carbon mineralization is shown to offer a promising approach for the permanent storage of CO_2, while transforming waste materials into usable resources. Integrated carbon capture and utilization (CCU) strategies further contribute to reducing emissions and replacing conventional raw inputs, thereby advancing industrial sustainability. Pyrolysis of lignocellulosic biomass is another efficient method for long-term carbon sequestration, producing biochar that stores

Table 1 Ethnobotanical contributions to eco-friendly alternatives and methane mitigation

Source	Study focus	Key findings	References
Plant-based alternatives	High-protein plant-based cheese	Developed using waxy starch, plant-protein isolate, and coconut oil; superior texture and meltability compared to commercial options	Dobson and Marangoni [15]
	Environmental impact of plant-based meat	Plant-based meat analogues reduce GHG emissions by 62% compared to beef burgers	Tang et al. [43, 44]
	Life cycle assessment: beef versus plant patties	Plant-based patties have significantly lower GHG emissions, water consumption, and land use	
	Sustainability of plant-based dairy	Offers 59–71% reduction in GHGs, less land and water use, maintains nutrition	Craig et al. [13]
Livestock methane mitigation	Feed-based methane mitigation	Low residual feed intake, diet quality, and concentrate-based rations reduce CH_4 per kg of milk	Broucek [8]
	Broad strategies for methane control	Multiple interventions (e.g., PSMs, feed additives, enzyme inhibitors); PSMs as a promising natural option	Alemneh and Getabalew [3]
	Factors influencing CH_4 production	Feed type, genotype, microbial profiles, forage selection, and stocking rate influence methane; a systems-based approach is needed	Galati et al. [18]

carbon over extended periods. Additionally, microbial conversion using microalgae has shown success in transforming CO_2 into biofuels and valuable bioproducts. These findings highlight the promise of combining various CO_2 capture technologies to cut emissions, generate value-added products, and deliver both economic and environmental benefits [40]. A study evaluated the carbon emission mitigation potential of selected tree species around three industrial sites: weaving, brick kiln, and cosmetic manufacturing. The findings indicated that *Moringa oleifera* exhibited the highest total biomass at inside-industry locations, with values of 2.58, 0.56, and 4.57 Mg ha^{-1} across the respective sites. Among the species, *Syzygium cumini* had the greatest total carbon stock and sequestration within the weaving (2.82 and 10.32 Mg ha^{-1}) and brick kiln industries (3.78 and 13.5 Mg ha^{-1}), while *Eucalyptus camaldulensis* led in the cosmetic industry (7.83 and 28.70 Mg ha^{-1}). Outside the industrial areas, carbon stock and sequestration were generally lower. Nevertheless, top performers included *Dalbergia sissoo* for the weaving industry (0.97 and 3.54 Mg ha^{-1}), *Moringa oleifera* at the brick kiln site (1.26 and 4.63 Mg ha^{-1}), and *Vachellia nilotica* at the cosmetic industry site (2.51 and 9.19 Mg ha^{-1}). Belowground carbon data indicated that soil carbon stock was highest in the top 0–20 cm layer and decreased with depth. This trend was consistent across all industries and both inside and outside zones. Overall, the findings confirm that urban and industrial-adjacent tree species, especially *Syzygium cumini*, *Eucalyptus camaldulensis*, and *Moringa oleifera*, play a significant role in carbon sequestration and can be strategically utilized in nature-based industrial emission mitigation strategies [51]. Another study revealed that the industrial sector contributes 21% of global CO_2 emissions, necessitating targeted carbon capture strategies due to the sector's process diversity. Unlike power generation, industrial facilities typically feature numerous smaller emission sources with diverse CO_2 concentrations, making carbon capture efforts more complex and cost-sensitive. In chemical industries, however, high-purity CO_2 streams present more affordable opportunities for early deployment of carbon capture and storage (CCS). Despite this, CCS cost estimates vary widely between sectors, hindering broader, sector-wide adoption. Additionally, policy challenges remain significant—global market competition can disadvantage early CCS adopters unless measures like carbon pricing or trade adjustments are established to level the playing field [24]. Another study assessed the importance and economic viability of CCS technologies in achieving substantial reductions in greenhouse gas emissions, with a particular focus on their application in the cement industry. Estimated costs for coal-fired post-combustion capture (PCC), natural gas-fired PCC, and Cryogenic Carbon Capture (CCC) were examined, accounting for regional variations in fuel and capital expenses. These figures were incorporated into the MIT Economic Projection and Policy Analysis (EPPA) model to forecast long-term deployment of industrial CCS technologies under alternative policy scenarios. The analysis demonstrated that industrial CCS remains a vital mitigation tool, particularly where negative emission technologies are lacking, to meet strict climate objectives. The modelling showed that CCS allows for the continued use of energy-intensive industrial products while delivering substantial emission cuts across industries and regions. The developed methodology also supports more comprehensive CCS planning and its integration into global energy and economic modelling structures [17]. Additionally, another study investigated the CO_2 capture capability of industrial by-products and urban waste, targeting carbon neutrality goals by 2050. Using CO_2 fixation experiments with constant flow aeration, the research assessed a range of by-products especially those high in free calcium from industrial waste disposal—for their carbon sequestration capacities. The concept of 'Disposal Facility CCS' was introduced, and estimates showed that such facilities can offer significant CO_2 capture potential, in some cases outperforming natural forest sequestration. Using national data, the authors also projected the capture potential of industrial by-products across Japan, supporting the vision of a future urban environment designed for passive carbon capture. These findings highlight the viability of integrating industrial waste into broader climate mitigation strategies [22]. A different study highlights the growing relevance of indoor plants in enhancing indoor air quality (IAQ), especially in the post-COVID-19 context, where people spend most of their time in enclosed spaces. Evidence supports that species like *Ficus benjamina*, *Chlorophytum comosum*, and *Dracaena* sp. significantly reduce airborne pollutants, offering a natural and cost-effective method to combat sick-building syndrome and respiratory ailments. The pollutant removal efficiency varies with plant species and specific traits such as leaf area, thickness, photosynthetic rate, and involvement of plant parts (leaves, roots, stomata). Despite their proven benefits, standardized criteria for selecting optimal indoor plants and understanding their influence on microclimatic factors (humidity, temperature, ventilation) remain lacking. The findings highlight the need for controlled, real-time indoor environment simulations to better quantify the biophilic and air-purifying roles of houseplants [38]. Air phytoremediation is emerging as an effective and economical alternative to traditional air purification methods, which typically demand significant investment in specialized infrastructure and technical skills. Beyond their ability to remove pollutants, certain plants also play a role in sequestering CO_2 through photosynthesis, supporting efforts to mitigate greenhouse gas emissions (Table 2). This approach benefits from the combined action of plants and the diverse microbial communities that inhabit leaf surfaces, making it a sustainable solution both environmentally and economically. The success and efficiency of phytoremediation techniques largely depend on the unique characteristics of plant species and the diversity of microbes present on the leaves. Consequently, targeted research is essential to identify and develop optimal plant-microbe partnerships for effective urban air quality management [9].

5 Traditional Medicine and Sustainable Resource Management

The rapid depletion of natural resources—driven by urban expansion, global warming, and habitat loss—poses a significant threat to the sustainability of traditional medicine systems that depend on herbs, minerals, and animal-based ingredients. The use of animal products in particular often raises ethical and ecological concerns. Ayurveda, however, exemplifies a sustainable approach by promoting non-lethal, respectful harvesting methods, such as collecting milk, honey, or wool in ways that do not harm animals or disrupt ecosystems. These practices embody an ethical and environmentally sound model that can be revisited and adapted for modern needs. By integrating traditional wisdom, contemporary strategies for resource use in traditional medicine can be reframed to prioritize both conservation and the enduring practice of these systems [37]. In addition, the practice of ingredient substitution in traditional Tibetan medicine stands out as a vital adaptive strategy within changing socio-ecological conditions. This approach not only demonstrates sophisticated ecological

Table 2 Ethnobotanical insights and technologies for industrial carbon capture

Aspect/study focus	Key findings	Industrial and environmental implications	References
Ethnobotanical knowledge	Identifies plant species with high carbon sequestration; traditional practices (agroforestry, mixed cropping, high-biomass species)	Informs afforestation, reforestation, and land management for industrial-scale carbon sequestration	Teffo et al. [46]
CO_2 capture technologies	Post-combustion and oxy-fuel combustion effective in cement industry; carbon mineralization for permanent storage; CCU strategies	Reduces emissions, enables waste-to-resource conversion, replaces raw materials, supports sustainability	Shammi and Akter [40]
Biomass pyrolysis	Pyrolysis of lignocellulosic biomass produces biochar for long-term carbon storage	Efficient carbon sequestration method for industrial application	Shammi and Akter [40]
Microbial carbon conversion	Microalgae used to transform CO_2 into biofuels and bioproducts	Promising for converting industrial CO_2 emissions into valuable products	
Tree species for industrial sites	*Moringa oleifera*, *Syzygium cumini*, *Eucalyptus camaldulensis*, *Dalbergia sissoo*, *Vachellia nilotica* excels at carbon sequestration	Strategic use in urban/ industrial zones for emission mitigation; species selection varies by industry type and location	Zubair et al. [51]
Soil carbon storage	Highest soil carbon stock in top 0–20 cm layer, decreasing with depth	Soil management critical for maximizing belowground carbon sequestration	Zubair et al. [51]
Industrial emissions and CCS	Industry responsible for 21% of global CO_2 emissions; multiple, diverse emission sources; high-purity CO_2 streams in some sectors	Targeted CCS strategies needed; cost and policy challenges for large-scale adoption	Leeson et al. [24]
CCS cost-effectiveness	CCS critical for deep emission cuts (especially in cement); cost varies by region and technology; enables continued industrial output	Informs policy and planning for integrating CCS in global industrial sectors	Farrell et al. [17]
Industrial by-products for CCS	Industrial by-products can fix CO_2; 'Disposal Facility CCS' concept; can outperform natural forests in some cases	Urban/industrial waste integration into climate mitigation; supports urban passive carbon capture vision	Komine et al. [22]

(continued)

Table 2 (continued)

Aspect/study focus	Key findings	Industrial and environmental implications	References
Indoor plants and air quality	*Ficus benjamina, Chlorophytum comosum, Dracaena* sp. reduce indoor pollutants; efficiency varies by species and traits	Natural, cost-effective method for improving indoor air quality; need for standardization and further research	Ravindra and Mor [38]
Air phytoremediation	Certain plants and phyllosphere microbes assimilate, degrade, or transform NO_2, SO_2, O_3, VOCs, PMs; also sequester CO_2 via photosynthesis	Low-cost, sustainable alternative for urban air quality management and greenhouse gas mitigation; requires targeted research	Chakravorty et al. [9]

understanding but also enables resilience in the face of resource scarcity and environmental changes. Through the lens of Resilience Theory, particularly concepts like adaptive cycles and panarchy, it becomes evident that traditional Tibetan medicine operates within a flexible, multi-scalar system that avoids rigidity traps. In contrast, the increasing industrialization and pharmaceutical standardization of traditional medicines impose inflexible frameworks that risk undermining sustainable resource use and cultural knowledge systems. Recognizing and institutionalizing the adaptive strategies inherent in traditional practices could support more resilient and sustainable medicinal production systems [19]. This study documented traditional ecological knowledge among residents of Chimalacatlán, focusing on local flora and fauna use and management. Using ethnographic surveys and chain-referral sampling, 48 individuals with extensive local knowledge were interviewed. Respondents identified 83 plant species and 44 animal species commonly used for food, medicine, fuel, ornamentation, and as pets. Key species identified for future sustainable management include *Odocoileus virginianus* (white-tailed deer) and *Ctenosaura pectinata* (Mexican spiny-tailed iguana) among animals, along with *Amphipterygium adstringens* and *Bursera linanoe* among plants. These findings highlight the significance of local knowledge in managing environmental resources and highlight the necessity for greater collaboration between communities and policymakers to ensure sustainable use and conservation efforts [20]. Traditional medicine systems, which are closely tied to natural ecosystems, exemplify the intrinsic relationship between environment, health, and culture. Deforestation poses a serious threat by diminishing biodiversity, undermining land rights, restricting access to resources, and jeopardizing cultural survival. Within this context, herbal medicine stands out as a sustainable practice that leverages locally available resources, thereby supporting ecological integrity, community well-being, and economic livelihoods. The study emphasizes the crucial role of herbal medicine in promoting sustainability, reinforcing the interconnectedness of nature, culture, and health [14]. The global rush for medicinal plants—which has intensified due to the COVID-19 pandemic, rising skepticism toward synthetic pharmaceuticals, and a renewed interest in natural remedies—has put considerable strain on wild plant populations. While these species offer immense therapeutic and economic value, unsustainable harvesting now threatens their survival and the ecosystems they support. Many medicinal plants are slow-growing, endemic, or keystone species, making them particularly vulnerable. Overharvesting, habitat loss, and lack of local awareness compound the problem. A sustainable future centers on integrated approaches that blend scientific innovation, policy development, cultivation standards, and traditional knowledge. Community involvement and transdisciplinary collaboration are key to ensuring conservation while meeting global health demands [5]. Over the past two decades, the global interest in traditional medicine, particularly herbal remedies, has grown substantially. Both developed and developing nations are experiencing significant economic benefits from the expanding herbal market. However, despite the popularity of these products, reports highlight inconsistencies in quality. The production of herbal medicines is complex, with quality often compromised during harvesting, storage, and processing. Adopting good agricultural and harvesting practices is crucial for ensuring the safety and quality of medicinal plants. Harvesting the correct plant part at its optimal developmental stage, maintaining proper drying and storage conditions, and utilizing efficient processing techniques are key to producing consistent, high-quality herbal products. When these practices are followed, the safety, efficacy, and reliability of herbal medicines can be assured, boosting market potential and contributing to sustainable growth in the herbal industry [31]. Herbal medicines have been integral to healthcare since the beginning of human civilization and remain widely used today. Their therapeutic, pharmacological, and economic significance continues to gain recognition, though the extent of this varies across countries. Despite the growing use of medicinal plants, there is a lack of structured legislative controls, which has hindered their integration into modern medicine. One of the major barriers to this integration is the absence of solid scientific and clinical evidence regarding the efficacy and safety of herbal products. While herbal medicine research has expanded globally, many clinical researchers and ethics committees still express concerns about addressing the associated ethical issues. This article focuses on the clinical, ethical, and regulatory challenges surrounding herbal medicine, highlighting the need for a more structured approach to ensure their safe and effective use in healthcare [33]. Medicinal plants are central to India's Ayurvedic healthcare tradition. While only 10% of plant constituents are known, the demand for herbal drugs has surged, putting pressure on natural resources. Ayurvedic harvesting methods, validated by modern science, emphasize timing based on plant parts, seasons, and lunar cycles to maximize therapeutic efficacy. Environmental factors such as climate, altitude, and cultivation practices also influence plant metabolites, directly impacting the effectiveness of medicinal plants. Proper collection and processing techniques are essential for maintaining drug potency and therapeutic value [45]. Indigenous Indian medicine relies heavily on medicinal plants, yet only a fraction of the species has been thoroughly studied. Of the 300,000 known plant species, only a small number provide well-defined drugs, and only about 10% of their organic constituents are understood. The active principles in medicinal plants are made available through proper harvesting

Table 3 Sustainability challenges and adaptive strategies in traditional medicine systems

Category	Key findings	References
Sustainable use of animal resources in Ayurveda	Ayurveda promotes ethical, non-lethal use of animals (milk, honey, wool) that does not harm biodiversity or ecosystems	Rastogi and Kaphle [37]
Adaptive strategies in Tibetan medicine	Ingredient substitution in Tibetan medicine is a flexible response to ecological change, reflecting resilience and deep ecological knowledge	Glover [19]
Herbal medicine as a sustainability model	Herbal practices support ecological balance, local economies, and health, tied closely to land rights and biodiversity	de Souza et al. [14]
Global demand and overharvesting of medicinal plants	COVID-19 and distrust of synthetics led to overharvesting; many medicinal species are slow-growing or keystone. Calls for integrated conservation methods	Alum [5]
Quality concerns in herbal medicine production	Despite popularity, quality is often compromised during collection and processing; good harvesting and storage practices are essential	Pandey and Savita [31]
Regulatory and ethical challenges	Lack of legislation and clinical evidence hinders the integration of herbal medicine into modern systems; ethical concerns remain	Parihar et al. [33]
Ayurvedic harvesting techniques	Traditional Ayurvedic harvesting is guided by timing, plant parts, and lunar cycles. Climate and altitude affect medicinal efficacy	Tavhare and Nishteswar [45]

methods. Ancient Ayurvedic practices, as outlined by Acharya Susruta and Charaka, emphasize the importance of factors like plant part, soil conditions, and the desired therapeutic effects. Modern scientific data supports these methods, highlighting the need for proper harvesting to preserve the plant's active compounds [35] (Table 3).

6 Policy and Industrial Integration

The circular economy is vital for sustainable development, aiming to decouple economic growth from resource depletion and environmental harm. To meet the EU Green Deal and net-zero objectives, the Visegrad (V4) countries, The Czech Republic, Hungary, Poland, and Slovakia, are implementing targeted strategies. A recent analysis examined the influence of the circular economy (CIR), green innovation (INV), renewable energy (REE), economic growth (GDP), and urbanisation (URB) on carbon emissions (CO_2) for the years 1990–2022 using CS-ARDL and quantile regression methods. The findings indicate that CIR, INV, and REE play a significant role in reducing CO_2 emissions, while GDP and URB contribute marginally to increased emissions [42]. The circular economy (CE) also delivers a transformative framework for sustainability within the tourism sector, fostering innovation and resource efficiency. In tourism and hospitality, research based on survey data from 180 professionals in Saudi Arabia and Egypt, analyzed using partial least squares structural equation modelling, reveals that CE encourages low-carbon behavior (LB), which in turn enhances environmental performance (EP). Moreover, eco-friendly behaviour (EB) further strengthens the connection between LB and EP, highlighting CE's influence on advancing sustainable industry practices and improving environmental results [50]. Additionally, policies promoting ethnobotanical approaches are critical for acknowledging indigenous knowledge, supporting sustainable resource use, and boosting biodiversity conservation. Collaboration among policymakers, industries, and indigenous communities enables better resource management, supports sustainable development, and ensures the inclusion of traditional knowledge within modern frameworks [23].

7 Challenges and Future Prospects

Ethnobotanical knowledge, while often grounded in empirical tradition, must undergo thorough scientific validation to establish the effectiveness of medicinal plants [41]. Around 80% of the world's population depends on plant-based medicines, highlighting the critical need to substantiate these practices with scientific evidence [36]. Historical records illustrate that numerous modern pharmaceuticals, including morphine and quinine, originated from ethnobotanical investigations, demonstrating the potential for discovering novel bioactive compounds [11, 36].

The commercial development of scientifically validated ethnobotanical products has the potential to create new markets for herbal medicines, generating benefits for both local communities and the broader global economy. However, ensuring quality control and standardization is essential for maintaining the safety and efficacy of these remedies, thereby fostering consumer confidence and wider acceptance in the marketplace [26]. High-amplitude ultrasonic processing has emerged as a promising technique for industrial processes in sectors such as food and beverage manufacturing. However, transitioning from laboratory to large-scale operations has been hindered by decreased ultrasonic intensity and trade-offs in product quality. The advent of barbell horn ultrasonic technology (BHUT) addresses this challenge, enabling high-amplitude ultrasonics at industrial scale and supporting scale-up factors exceeding 50 without compromising product quality. This development positions BHUT as a competitive alternative for commercial production [34]. In the context of industrial reliability, traditional testing methods encounter obstacles, particularly as comprehensive assessments of security and software reliability become increasingly vital. Furthermore, discrepancies between interconnect and gate delay reductions continue to challenge further scaling in industrial applications [39]. Looking ahead, ethnobotanical research is increasingly integrating technological innovations such as advanced databases, mobile applications, machine-learning techniques, Geographic Information Systems (GIS), remote sensing and big data analytics. These tools greatly enhance data collection, management, and analysis, while assisting researchers in addressing issues of ethics and cultural sensitivity [6]. At the same time, the field is sharpening its emphasis on sustainable development and biodiversity conservation by integrating traditional indigenous wisdom with modern scientific methods. This fusion aims to facilitate the discovery of new drugs and refine conservation strategies, directly responding to pressing concerns about phyto-diversity loss and ecological sustainability in a rapidly evolving global landscape [32].

8 Conclusion

The traditional ethnobotanical knowledge of the indigenous people, based on the well-established values of preservation of resources, ecosystem stability, and conservation of biodiversity, is needed to achieve sustainable development and a carbon-neutral future. The Indigenous peoples as stewards of this knowledge can offer very important information concerning low-impact plant-based industries, sustainable harvesting, and climate-sensitive land use. All these insights are in line with international agendas such as improving health, alleviating poverty, responsible production, climate change mitigation and conservation of terrestrial ecology. By integrating such traditional ecological knowledge into the industrial structures, natural carbon sequestration strategies can be enhanced in both rural and urban settings, the circular bioeconomy practices can be promoted, and the materials with high emissions can be replaced with renewable florals. The combination of the knowledge that has been acquired throughout the centuries with the new achievements of modern technologies offers an innovative approach to enhancing the sustainability of industries, making the communities more conservation-focused, and boosting the health of livestock with the help of plant-based therapies. Nevertheless, robust policy support, scientific justification of adaptation, and ethical systems that ensure recognition, equality and fair sharing of benefits with Indigenous knowledge holders are all required to be successfully implemented. Ethnobotany has the potential to evolve from a dismissed object of culture to a scientifically sound means of climate resilience and environmental sustainability due to the establishment of goodwill and the recognition of Indigenous intellectual property.

References

1. Al Khaffaf I, Tamimi A, Ahmed V (2024) Pathways to carbon neutrality: a review of strategies and technologies across sectors. Energies 17(23):6129
2. Albrecht A, Kandji ST (2003) Carbon sequestration in tropical agroforestry systems. Agric Ecosyst Environ 99(1–3):15–27
3. Alemneh T, Getabalew M (2019) Strategies to reduce methane emission in ruminants. Int J Ecol Ecosolution 6:16–22
4. Alexiades MN (2003) Ethnobotany in the third millennium: expectations and unresolved issues. Delpinoa 45(1):15–28
5. Alum EU (2025) Sustainable harvesting of medicinal plants: balancing therapeutic benefits with environmental conservation. Agroecol Sustain Food Syst 49(3):380–385
6. Banisetti DK, Kosuri NP (2023) Ethnobotanical research in the digital age: harnessing technology for data collection and analysis. Int J Indig Herbs Drugs 1–6
7. Baral A (2014) Demystifying the carbon neutrality of biomass. The International Council on Clean Transportation (ICCT)
8. Broucek J (2018) Methane abatement strategies based on genetics and dietary manipulation of ruminants: a review. Arch Zootec 67(259):448–458
9. Chakravorty A, Raghavan V, Grace AN, Kim SY, Van Le Q (2022) Plants and microorganisms for phytoremediation of air. In: Current developments in biotechnology and bioengineering. Elsevier, pp 109–131
10. Chandrashekara UM, Sankar SJFE (1998) Ecology and management of sacred groves in Kerala, India. For Ecol Manag 112(1–2):165–177
11. Colombo ML, Dalfrà S, Scarpa B (2012) Scientific evidence of ethnobotanical and Mediterranean knowledge of food and well-being plants. J Pharm Sci Res 4(1):1662
12. Correa CM (2001) Traditional knowledge and intellectual property, vol 17. The Quaker United Nations Office (QUNO), Geneva. ISBN 2-9700323–0-9
13. Craig WJ, Messina V, Rowland I, Frankowska A, Bradbury J, Smetana S, Medici E (2023) Plant-based dairy alternatives contribute to a healthy and sustainable diet. Nutrients 15(15):3393
14. De Souza A, Ferreira P, Hüther C, de Jesus I, de Carvalho A, de Oliveira RPRF, Ferreira V (2024) The sustainable power of herbal medicine: bridging tradition and ecology, pp 1–12
15. Dobson S, Marangoni AG (2023) Methodology and development of a high-protein plant-based cheese alternative. Curr Res Food Sci 7:100632
16. Eduard C (2023) Conservation of biodiversity through ethnobotany and traditional knowledge. J Biodivers Bioprospecting Dev 9:26
17. Farrell J, Morris J, Kheshgi H, Thomann H, Paltsev S, Herzog H (2018) The role of industrial carbon capture and storage (CCS) in emission mitigation. In: 14th greenhouse gas control technologies conference Melbourne, pp 21–26
18. Galati RL, Silva LCRP, Fonseca ASR, da Silva Cabral L, Cerom MFSQ (2023) Challenges and strategies for reducing enteric methane emissions in sustainable animal production. Concilium 23(14):301–316
19. Glover DM (2021) Traditional medicines in a global economy: resource sustainability and resilience in the traditional Tibetan medical practice of ingredient substitution. Hum Ecol 49(1):33–42
20. Hernández-Tapia R, Valverde T, Aranda A, Martínez-Peralta C, Platas-Neri D (2018) Traditional knowledge as a basis for the development of a sustainable resource management program. Ethnobiol Lett 9(2):1–11
21. Kandari LS, Bisht VK, Bhardwaj M, Thakur AK (2014) Conservation and management of sacred groves, myths and beliefs of tribal communities: a case study from North-India. Environ Syst Res 3:1–10
22. Komine H, Saitou Y, Suzuki K (2024) Carbon dioxide capture potential of industrial by-products for post-carbon society. Environ Geotech 11(1):1–13

23. Kumar A, Kumar S, Komal, Ramchiary N, Singh P (2021) Role of traditional ethnobotanical knowledge and indigenous communities in achieving sustainable development goals. Sustainability 13(6):3062
24. Leeson D, Ramirez A, Mac Dowell N (2019) Carbon capture and storage from industrial sources. In Bui M, Mac Dowell N (eds) Magnesium batteries: research and applications. Royal Society of Chemistry, pp 296–314
25. Mercader J (2009) Mozambican grass seed consumption during the Middle Stone Age. Science 326(5960):1680–1683
26. Mukherjee PK, Harwansh RK, Bahadur S, Banerjee S, Kar A (2016) Evidence based validation of Indian traditional medicine–way forward. World J Tradit Chin Med 2(1):48–61
27. Murthy IK, Gupta M, Tomar S, Munsi M, Tiwari R, Hegde GT, Ravindranath NH (2013) Carbon sequestration potential of agroforestry systems in India. J Earth Sci Clim Chang 4(1):1–7
28. Nair PKR, Kumar MB, Nair VD (2009) Agroforestry as a strategy for carbon sequestration. J Plant Nutr Soil Sci 172(1):10–23
29. Negi CS (2005) Religion and biodiversity conservation: not a mere analogy. Int J Biodivers Sci Manag 1(2):85–96
30. Ormsby AA (2011) The impacts of global and national policy on the management and conservation of sacred groves of India. Hum Ecol 39(6):783–793
31. Pandey AK, Savita R (2017) Harvesting and post-harvest processing of medicinal plants: problems and prospects. Pharma Innov J 6(12):229–235
32. Pandey AK, Tripathi YC (2017) Ethnobotany and its relevance in contemporary research. J Med Plants Stud 5(3):123–129
33. Parihar N, Barve K, Bhatt LK (2023) Clinical, ethical, and regulatory issues of botanicals. In: Phytopharmaceuticals and herbal drugs. Academic Press, pp 265–279
34. Peshkovsky AS (2017) From research to production: overcoming scale-up limitations of ultrasonic processing. In: Ultrasound: advances for food processing and preservation. Academic Press, pp 409–423
35. Pushpan R, Kumari H, Nishteswar K (2012) Influence of harvesting and processing methods on the active principles of medicinal plants. Int J Pharm Biol Arch 3:1283–1286
36. Qureshi R, Ghazanfar SA, Obied H, Vasileva V, Tariq MA (2016) Ethnobotany: a living science for alleviating human suffering. Evid-Based Complement Altern Med eCAM 2016:9641692
37. Rastogi S, Kaphle K (2011) Sustainable traditional medicine: taking the inspirations from ancient veterinary science. Evid-Based Complement Altern Med 2011(1):151435
38. Ravindra K, Mor S (2022) Phytoremediation potential of indoor plants in reducing air pollutants. Front Sustain Cities 4:1039710
39. Seshan K (2018) Reliability issues: reliability imposed limits to scaling. In: Handbook of thin film deposition. William Andrew Publishing, pp 43–62
40. Shammi M, Akter B (2024) CO_2 sequestration for conventional utilization and industrial application. In: Decarbonization strategies and drivers to achieve carbon neutrality for sustainability. Elsevier, pp 299–319
41. Subba B, Srivastav C, Kandel RC (2016) Scientific validation of medicinal plants used by Yakkha community of Chanuwa VDC, Dhankuta, Nepal. Springerplus 5:1–14
42. Takyi KN, Gavurova B, Ofori C, Mikeska M, Sampene AK (2024) Assessing the role of circular economy and green innovation in mitigating carbon emissions in the Visegrad countries. Int J Renew Energy Dev 13(6):1149–1161
43. Tang M, Miri T, Soltani F, Onyeaka H, Al-Sharify ZT (2024) Plant-powered plates: a life cycle assessment of plant-based meats in tackling climate change. Preprints. https://doi.org/10.20944/preprints202404.1262.v1
44. Tang M, Miri T, Soltani F, Onyeaka H, Al-Sharify ZT (2024) Life cycle assessment of plant-based vs. beef burgers: a case study in the UK. Sustainability 16(11):4417
45. Tavhare SD, Nishteswar K (2014) Collection practices of medicinal plants-vedic, ayurvedic and modern perspectives. Int J Pharm Biol Sci 5:54–61
46. Teffo TK, Ramalepe P, Risenga I (2022) A short communication on the ethnobotany, phytochemistry, pharmacological evidence and ecosystem restoration potential of South African Portulacaria afra. J Med Plants Stud 10:112–115

47. Vipat A, Bharucha E (2014) Sacred groves: the consequence of traditional management. J Anthropol 2014(1):595314
48. Wadley L, Sievers C, Bamford M, Goldberg P, Berna F, Miller C (2011) Middle stone age bedding construction and settlement patterns at Sibudu, South Africa. Science 334(6061):1388–1391
49. Wang F, Harindintwali JD, Yuan Z, Wang M, Wang F, Li S, Chen JM (2021) Technologies and perspectives for achieving carbon neutrality. The Innovation 2(4):1–22
50. Zaki K (2024) The impact of circular economy on environmental performance in the tourism and hospitality industry: the role of low-carbon behavior and eco-friendly behavior. Int J Tour Hosp Stud 7(2):285–304
51. Zubair M, Yasin G, Qazlbash SK, Ul Haq A, Jamil A, Yaseen M, Guo W (2022) Carbon sequestration by native tree species around the industrial areas of Southern Punjab, Pakistan. Land 11(9):1577

Reducing Carbon Footprint in the Construction Industry: Strategies from Design to Operation

Shrikant M. Harle, A. B. Ranit, Swati Ambadkar, and S. Bejalwar

Abstract Various sources of carbon emissions of the construction lifecycle such as design, manufacture in the production of the construction materials, transportation, building, operations and maintenance, and lastly deconstruction are presented in the review study. Among the options that have been studied currently on how to reduce carbon footprints at different stages are the use of alternative sources of water, optimizations in the design, recycling construction waste, alternate additives to the materials and the basic enhancements in water and building system efficiency. Research has indicated that other materials, substances, procedures and systems could cut the amount of CO_2 emission by almost 90% at various stages of construction and operation of a building making construction industry to be the largest emitters of carbon of all establishments with minimal space required to cut down their emissions. This is a priceless tool as clients and other stakeholders can use it to make informed judgments on the choice of materials and systems and create environmental awareness on impacts of manufacturing, transporting, and constructing actions.

Keywords Construction industry · Carbon footprint · Carbon emissions · Sustainability · Environmental impact · Building materials

S. M. Harle (✉) · A. B. Ranit
Department of Civil Engineering, Prof Ram Meghe College of Engineering and Management, Badnera, Maharashtra, India
e-mail: shrikant.harle@prmceam.ac.in

S. Ambadkar
G H Raisoni University, Amravati, India

S. Bejalwar
Prof Ram Meghe College of Engineering and Management, Badnera, Maharashtra, India

S. S. Muthu (ed.), *Reduction of Industrial Carbon Footprint*, Environmental Footprints and Eco-design of Products and Processes,
https://doi.org/10.1007/978-3-032-10913-2_4

1 Introduction

Although global analysis of the construction industry is a subject of controversy, the industry is still being scrutinized due to its subjective responsibility and the contribution made to global carbon dioxide (CO_2) emissions which roughly amounts to about 38%. The construction business places the urgent need on the necessity to intervene on the climate change and decrease greenhouse gas (GHG) emissions. These are direct and indirect-based emissions which are characterized as embodied carbon and operational carbon. As the rate of urbanization grows and with a subsequent growth of the demand in the infrastructure, with no active measures, the carbon footprint of the construction activity is expected to grow. In order to develop sufficient measures of cutting the carbon footprint of the buildings and structures, it is beneficial to grasp the full scale of the emissions in the life cycle of buildings [11].

The embodied carbon will include the total amount of greenhouse emissions generated through mining, processing, delivering, and putting the building materials altogether. This includes mining of natural resources like limestone or iron ore, production of cement and steel, production of building products and transportation of this construction material into the construction sites. They begin well in advance before a building is even factoring its occupancy and represent a huge proportion of the total number of building lifetimes emissions. As an illustration, cement manufacturing alone accounts to approximately 7–8% of the global total emissions in terms of carbon dioxide because the calcination process is also included during their manufacture and fossil-based sources of fuel are used as well. One is that embodied carbon has the potential to be reduced by the use of low-carbon materials, further optimization of even the structure, and the encouragement of using recycled material or locally produced materials along with efficient workmanship [13].

The operation carbon on the other hand is the emissions that are produced during the life period of the buildings use. This is what is employed to heat, ventilate, air condition (HVAC), light, and the appliances and water heating energy. The type of design, insulation, orientation, materials and source of energy to be used in the building directly affects these emissions. Historically, operational emissions make up most of a buildings total savings through its lifetime. In the case of improvements in energy efficiently and a growing use of renewable energy, however, embodied emissions are gaining relative importance. Thus, operational carbon minimization necessitates combinational efforts between passive design, high efficiency systems, intelligent controls and renewable sources of energy like solar pv or geothermal heating [40].

To fully mitigate the carbon emission caused by the built environment, there is a necessity to look to adopting whole-life carbon evaluation strategies aimed at trying to reduce the embodied and operational carbon of the building through phases of design and construction, use, maintenance and demolition or reuse. An early design is a determining factor because it determines the context of energy carrying and material selection. Application of the life cycle assessment (LCA) tools during planning is appropriate in order to help the stakeholders make an informed decision based on the consideration of performance, cost, and sustainability. The use of the modular constructions, the introduction of 3D printers, and green roofs can also be used as other technologies contributing to low emissions but should be used cautiously.

In addition, there is a necessity of industry collaboration and policy-level interventions in order to realize that the carbon reduction could be scaled up. Another positive input that can be made by the regulatory agencies and government is on the incorporation of the green build codes, support of low-carbon technology and the initiation of environmental reporting on the large civil works that the governments have. Industry partnerships, knowledge sharing, and platforms will also speed up the transfer to sustainable ways of construction. Awareness and accountability, likewise, are categorized under the certification systems, such as LEED, BREEAM, and IGBC, which influence developers to exceed their green threshold [36].

Carbon Reduction in construction is mentioned in Fig. 1. The hybridised methodology on reduction of carbon emission in the construction industry is the one that takes into account a number of issues which are cross-sectoral. It is not merely a question of technological creativity and the economic efficiency with which materials are utilised but also it deals with behaviour and education and policy change. The embodied and operational carbon allow the stakeholders to reduce the impact on the environment dramatically in terms of both constructions and infrastructures. The chapter may be utilized in highlighting the importance of the individual construction stages and present operational solutions of minimizing the carbon footprints which could eventually support the goals of resilience and sustainability developments across the globe with regard to climate resilient and sustainable development goals.

Roadmap

This chapter discusses the issue of the carbon footprint mitigation in the building sector in the context of the entire life-cycle of buildings, encompassing their design and specification, material production, transportation, assembly, use, service and deconstruction or re-use. It focuses on incorporating new digital design technologies, low-carbon and recycled materials, efficient supply chains, prefabrication, renewable energy, smart technologies and retrofit as main drivers of the emission reductions. Cross-cutting policies related to life cycle assessment, carbon capture and storage, optimization of improving the nexus of water and energy, or sustainable waste are also discussed together with the innovation of construction materials and global green building guidelines. The chapter will use targeted case studies, and as an example, India, to show how policy measures, academic participation and industry partnership can make the transition to low-carbon construction more rapid. In conclusion, the roadmap proposed in this paper is a way forward to take the construction industry out of its current status of being a significant source of emissions in the world and into the climatically resilient and sustainable development industry.

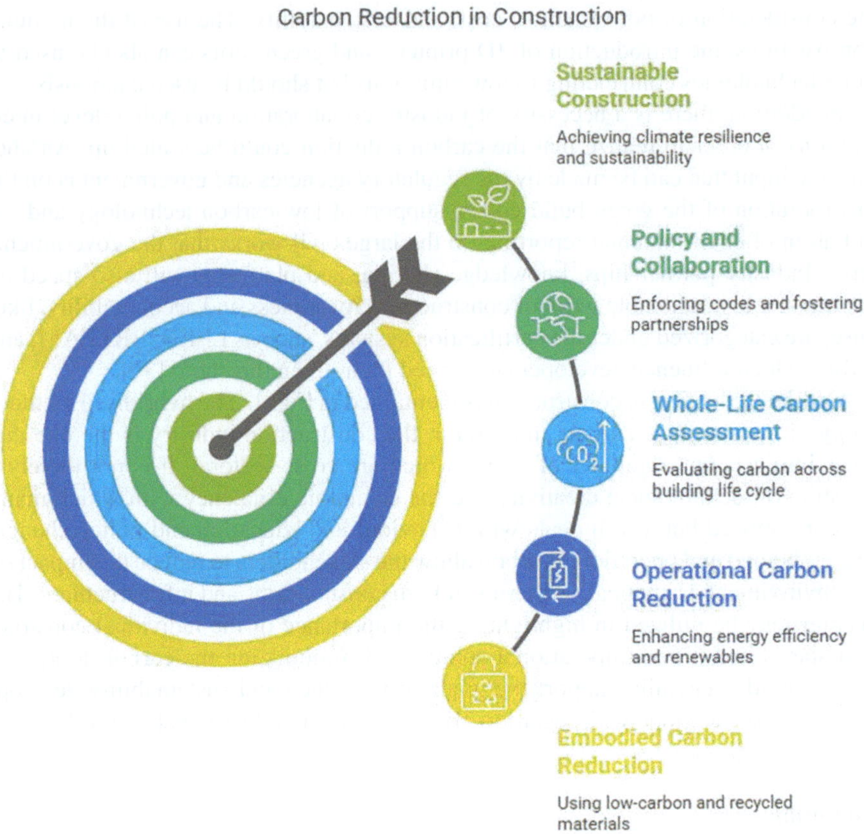

Fig. 1 Carbon reduction in construction

2 Carbon Footprint in the Construction Lifecycle

2.1 Design Phase

Architectural and structural design decisions are becoming central concerns in determining the long term eco performance and carbon performance of a building. At the beginning of any construction project that remains at a conceptual stage, the design orientation, layout planning, the material and the form of constructions that will be adopted by architects and engineers will significantly influence embodied and operational carbon footprints. An example is that an excessive proportion of the utilisation of a carbon-intensive material (user-reinforced concrete and structural steel) can cause a greatly magnified embodied carbon profile. Structural performance requires these kinds of resource materials in abundance and it is essential but as much as may be, the resources ought to be applied with moderation and modest alternatives.

Moreover, design solutions which fail to maximise the natural daylight or ventilation create an unnecessary over-dependence on artificial light and mechanical HVAC systems resulting in excess carbon to the lifecycle of buildings [19].

In order to overcome these challenges, construction industry tends to adopt digital design tools and simulation software that enables decision-making using data. The emergence of Building Information Modeling (BIM) system has made integrated design practices where collaboration between architectural designers, engineers, and contractors occurs on a seamless basis. Using BIM, the entire building can be modelled virtually to allow teams to see areas of inefficiency and optimize their structural and material selection to reduce waste and increase building energy performance. Furthermore, BIM is used to coordinate disciplines and early resolution of design conflicts, thus reducing costly, rework and wasted on-site materials [23].

Simultaneously, a new and essential tool maintains its environmental ranking on design options of a building over time LCA platforms provide the evaluation of the influence of design options on structure efficiency. They are tools to consider many possibilities of material use, construction techniques, energy systems, and end-of-life aspects giving a comparative review of different design possibilities. By utilizing LCA, the decision-makers can prioritize the low-carbon materials used, suggest using the modular/prefabricated parts, as well as anticipate a later reuse or rather repair, thus helping to reduce the embodied carbon footprint [17].

Energy modeling and simulation programs like EnergyPlus go further in making the design teams more capable in guessing and modeling how a certain building will respond to different climate conditions and through different operational constructions. These devices model heating and cooling loads, day lighting efficiency, thermal comfort and energy usage profiles. Through these repetitions of various design possibilities the engineers are able to dial in the thermal envelope, orientation, glazing ratios and shading devices to reach the most energy efficient arrangement thereby minimizing operational emissions. These sorts of simulations are of great value in determining the intent of the design and the reality of the outcome of the design behavior, especially in climate-responsive design [14].

Sustainable design process is mentioned in Fig. 2. In addition to digital tools, passive design integration is one of the approaches that can offer the most value in terms of emissions reduction related to a building. These methods make use of on-site natural resources and climates in controlling indoor conditions. Buildings can be oriented so as to utilize day lighting without heat gain, cross ventilation, use of eaves or louvers to provide shade, and inclusion of thermal mass to moderate indoor temperature levels so that mechanical systems are utilized much less. Passive design, done intelligently, can result in significant energy savings and, in addition, can provide greater comfort to occupants as well as better air quality in buildings [43].

Together these methods are a complete strategy of sustainable architectural and structural design: they include the digital analysis tools and LCA analysis as well as active principles of passive design. Not only can they decrease emissions and operation costs but also coincide with world environmental aims and construction building

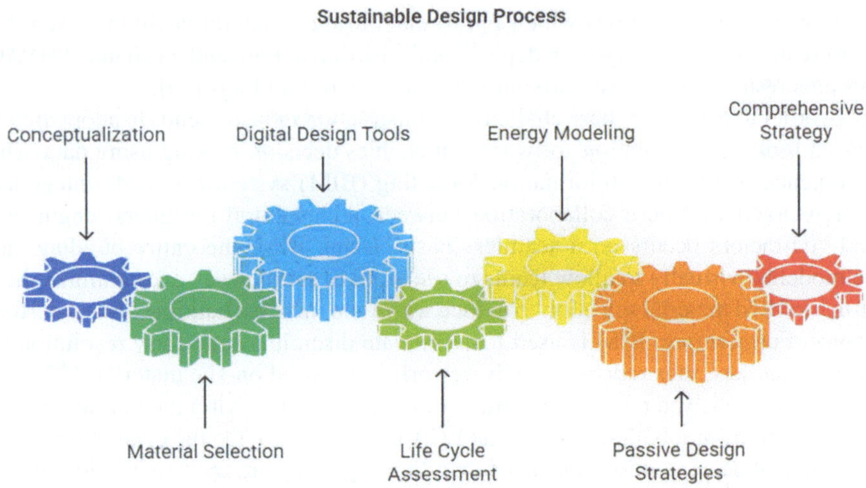

Fig. 2 Sustainable design process

certification. Such design choices eventually build strong future-grown buildings that are practical and environment friendly structures [6].

2.2 *Material Manufacturing*

The construction lifecycle is dependent on the material manufacturing, one of the processes that contributes significantly to the total carbon emissions of the industry. The stage entails stripping of raw materials, processing and transforming the materials into forms that can be used by the construction products. Cement is categorized as the most carbon intensive amongst these, mostly due to the manufacturing of basic Portland Cement (OPC). Procurement of OPC entails calcination process by which limestone is heated to a high temperature which besides heavy consumption of fossil fuels, emits huge amounts of carbon dioxide as by-products of OPC chemical reactions. Therefore, cement manufacturing alone is estimated to contribute approximately 7–8% of total global emission of CO_2 and this elevates the area as a point of decarbonization in the industry. Moreover, raw materials, including steel and aluminum, are not free of carbon footprint because the processes involving mining, smelting, and refining cannot be regarded energy-intensive and thus fall victim to high demands on the thermo energy and lead to record-breaking emissions [44].

Knowing that it is required to decrease the emissions of this type, the industry started exploring the low-carbon options and more actively transitioning to them. Some of the best ones can be attributed to the fact that OPC has been substituted with Supplementary Cementitious Materials (SCMs). The by-products of the industry usually take the form of fly ash, Ground Granulated Blast Furnace Slag (GGBS) and silica fume and even though they just increase the clinker content of the concrete, they bring about an increment in the long term performance and strength of the concrete structure. These substitutes have twofold advantages of reducing carbon emissions generated during the manufacture of cement and ensuring a productive manner of disposing of waste products that would otherwise lead to environmental pollution. Such absorption of SCMs by concrete mixes can lead to significant reductions in embodied carbon in the industry which, in turn, will enhance material performance [1].

Construction material comparison is observed in Fig. 3. Alongside replacement to cement, biogenic products have formerly appealed commercial entrants, including engineered timber and bamboo, among others on account of the low effects on the environment and renewability. There are structural products available such as Engineered Wood Products and in this case specifically Cross-Laminated Timber (CLT) that offer similar structural strength to any other traditional building material such as steel and concrete, however in a product with an extremely low carbon footprint. As the tree grows, the products sequester the carbon and retain it throughout the useful life of the building playing the role of long-term storage of carbon. Similarly, bamboo (fast regrowth rate belt) and high tensile strength has of late emerged as an alternative alternative in the space of modular or pre-fabricated constructions. Responsibly harvested, engineered Bamboo will provide us with sustainable yet lightweight viable alternative in many applications structural and non-structural [28].

The most important technological revolution in production of material has been the introdution of geopliver concrete which can possibly provide an alternative concrete free of cement. Geopolymer concrete occurs as a result of a chemical reaction between aluminosilicate rich industrial wastes e.g. fly ash or blast furnace slag, and alkaline solutions. It is a non-clinker process and results in an economically attractive (potentially 80–90% lower CO_2 emissions) concrete product with a binder having comparative or even superior mechanical properties than OPC based concretes. Besides reducing carbon, geopolymer concrete has proved highly resistant to chemicals,

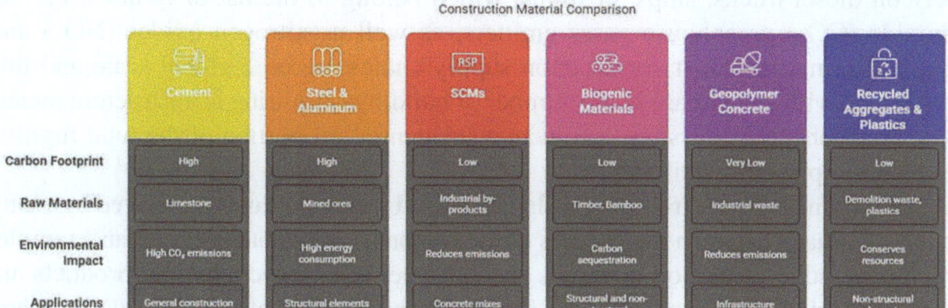

Fig. 3 Construction material comparison

thermal effects, and erosion by the environment; therefore, it holds potential in terms of infrastructure construction in hostile conditions [16].

Recycled aggregate and plastic innovations are also helping to ensure the sustainable production of materials. Demolition aggregates (RCA) may be used in fresh concrete mixes as a replacement to natural aggregates, decreasing the number of virgin materials and emission caused by quarrying processes. When re-processed and engineered into composite materials, recycled plastics create light, long-lasting, and corrosion resistant alternatives to non structural applications such as panels, piping and cladding. The effect of such practices is not only a saving of natural resources but also less strain on landfills and reconstitution of the loop in the lives of materials [49].

Overall, the manufacturing of material is a stage of the construction process that causes a large impact to achieve a high level of carbon savings through innovation, replacement, and recycling. The industry can also significantly reduce the environmental footprint by switching to new materials and materials that use natural sources away from conventional high carbon sources. With the rising demand in sustainable construction, it will be dependent on investments in further research, through policy support, and by active dissemination in the market in the use of environmentally friendly building materials, to allow material manufacturing to become a pillar of climate resilient infrastructure.

2.3 *Transportation*

Transportation—another often overlooked part of the construction lifecycle, can amount to a large percentage of total embodied carbon emissions of a project. The step includes transporting of the raw materials to the manufacturing and processing plants and then to the construction site after it is extracted at the source (which could be quarry, forest or mine). Besides, delivery of prefabricated elements, equipment, and machine parts also leads to transport emission. The majority of such movements rely on diesel trucks, ships, or trains, which belong to the list of frequent carbon dioxide (CO_2) greenhouse gases emitters, as well as nitrogen oxides (NO_x) and particulate matter. Most construction supply chains are on a global scale and this combined with an increase in the number of buildings requiring infrastructure means that transportation emissions can form a substantial proportion of the total fugitive carbon footprint of a building [2].

The strain on the environment is also increased where products are moved by a long haul or perhaps between the borders of the region or even countries. As an example, high demand construction materials such as steel or engineered wood products are often imported or may prefer to be imported on the basis of price or quality when local substitute products exist. Though the process of prefabrication has a potential to minimize waste and shorten time on-site, it is likely to contribute to the transportation emissions unintentionally, which can be significant when the manufacturing hubs are located a long way away from the project site. These logistical realities indicate that, when planning a project it is necessary to carefully analyze the transportation routing, transportation mode and distances when purchasing project procurement [47].

Various effective and easy solutions can be incorporated to ensure that any carbon emission impact is diminished during this stage. Possibly the easiest of the solutions would be to prioritize the usage of locally produced materials. Purchasing the supplies that are produced by local suppliers shortens transportation distances, fuel consumption as well as emissions exponentially. Also, it supports economic growth across a region and enhances supply chain resilience. Substitution of high-carbon materials by substitutes available locally where possible will enhance these advantages. As another example local supply of fly ash can reduce emissions, as well as lower the costs as compared to importing cementious materials [24].

The other worthy technique is in logistics which is optimisation by organizing and combining deliveries. Improved construction managers need not accept numerous smaller shipments on varied days or vendors to make bulk deliveries to limit repeated deliveries during the project. This will limit the number of trips, fill the vehicles to the maximum capacity, and waste less time at the destination areas, all of which lead to saving fuel and emissions. In addition, digital logistics management tools can be used to control the planning of deliveries, minimize delays and eliminate duplication of transport processes [3].

Green technologies also play a factor in adoption of green technologies in transportation so far as it helps to cut environmental implications. Electric or hybrid transport vehicles can be used in shorter- to medium-haul transportation and this can contribute significantly to the reduction of tailpipe emissions especially in construction locations where pollution already runs rampant in urban centers or near-urban fringe locations. Their carbon intensity is also reduced in case the vehicles operate on renewable energy sources. Although such cars or charging mechanism may be initially costly to invest in, the impact would be gigantic on the environment in the long run as well as operating costs. We should also where possible use rail and water transportation which offers lower emissions per ton-kilometer as compared to road freight where bulk material has to be moved as much as possible [38].

To sum up, the transportation aspect of the construction supply chain should also be addressed to minimize the embodied carbon emissions and enhance the sustainability in general. Three of the key themes on which project developers can play a substantive role in setting up low-carbon constructions are in giving attention to local sourcing, efficient logistics, and cleaner transport technologies. These interventions cannot only limit the emissions but increase the efficiency of the project, cost-effective nature, and environmental responsibility, the main pillars of contemporary sustainable construction practices. Strategies for sustainable construction transportation is mentioned in Fig. 4.

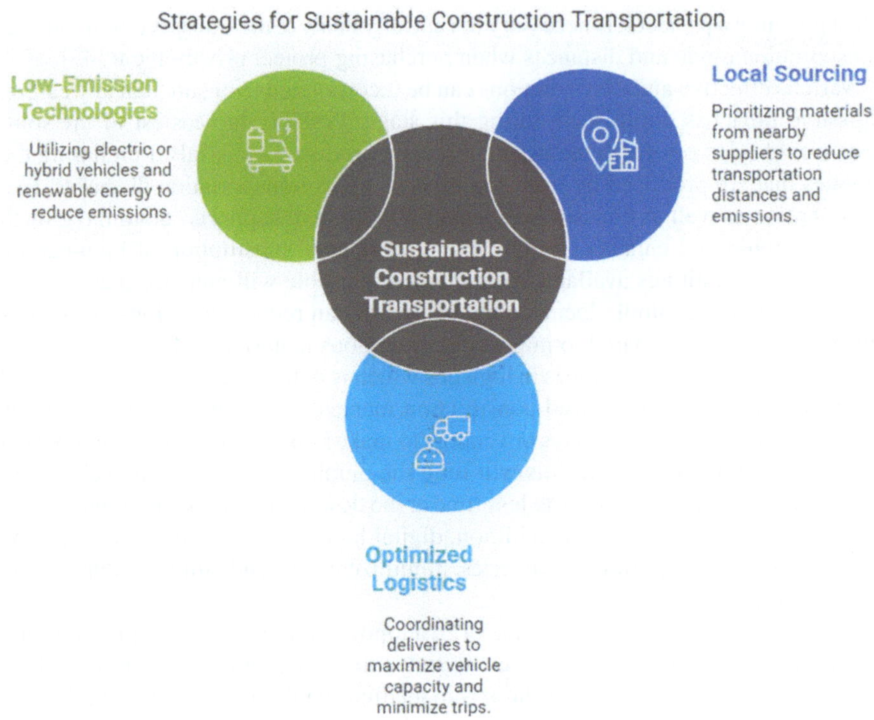

Fig. 4 Strategies for sustainable construction transportation

2.4 Construction Phase

At this stage of the constructing, one of the peak stages of carbon-footprint of the constructing development occurs as a great portion gets utilized at this stage. Intensive use of heavy machinery and equipment and diesel-powered parts, that is, excavators, bulldozers, cranes, or concrete mixers, is the primary source of such emissions. The machines as well consume a lot of fossil fuels in form of too much carbon dioxide (CO_2), nitrogen oxides (NO_x), and particulate matter that exacerbate the condition of global warming and local air pollution respectively. The emission of these pollutants not only amplifies climate change but it also causes health hazards especially to the construction workers and residents in the locality since it increases the severity of lung and respiratory complications in addition to other health challenges. As the impacts are so large, it is very important to discuss the effective strategies in order to enhance efficiency on-site and cut down emissions in this eminent phase [9].

Electrification of construction machines and gear is one of the most promising solutions that can resolve these challenges in the realm of environmental concerns. The no on-site output of electric powered equipment cannot break the carbon footprint of the construction increased by long distances. Furthermore, the electrical machines have reduced noise as opposed to their counterparts, diesel making them suitable in developing cities where noise pollution is an issue. Paying upfront costs in electric equipment may be more than that in conventional engine machines, but in the long run, low consumption of fuel, minimal expenses in maintenance and high sustainability benefits complement initial expenses impact and so on. The construction industry will be all the better placed to make this transition to more environmental friendly technologies with advancements in the availability of electric machinery, and better infrastructure in terms of charging points [45].

Introduction of on-site renewable energy to drive construction processes will be another good measure. An example will be, solar powered lights and, equipment charging points which can drastically minimize the use of diesel generators that are normally in operation at construction sites particularly those situated at remote or off grid places. There are ways to mitigate this by utilizing temporary construction offices, and lighting towers, as well as smaller tools that run off of solar energy- and by doing so, construction projects can find a way to lessen their reliance on fossil fuels and as a result, lessen the emissions. The practice further augers well in construction projects as it offers clean renewable energy that is sustainable considering that this can be unleashed even in places that are not connected to other power grids [20].

The other innovative idea becoming popular in the construction sector is the employment of prefabrication and modular construction methods. These techniques include the production of building materials in factory environments under controlled conditions which means more precision and efficiency of materials. Prefabrication decreases costs not only because of the waste of materials but also because of the minimization of energy that goes into the process. After the components are manufactured they are brought to the construction location to be assembled, which minimizes labor to be performed on-site, the use of heavy machinery and the time to be spent on the construction. This lowers the number of emissions of the machinery on-site, and transportation and give a general overall decrease in carbon footprint of the project. In addition, prefabrication eliminates the necessity of repeated supply of materials to the site thereby cutting down transportation emissions as well [21].

Reducing carbon emissions in construction phase can be observed in Fig. 5. Efficiency in the usage of materials is also a part of the sustainable approaches that should be implemented into the construction process. The transportation emissions are avoided, and the carbon scheme of the development is lower by the use of locally sourced materials. Also, the environmental effect of any construction could be further mitigated by the choice of lower embarked materials, sustainably sourced timber, recycled steel, or low-carbon concrete equivalent. It is also possible to introduce lean construction techniques that involve reducing waste, maximizing the use of resources, and increasing the efficiency of the overall project to construction teams. These solutions can assist in the minimization of useless material consumption, and the burden of construction on the environment [1].

Last but not least, the enhancement in construction sites management and operating culture can add a major impact in emission reductions. Options like streamlining the car path, vehicle idling hours, and introduction of a practicable logistically system can keep fuel usage and transportation emissions down on-site. The appropriate maintenance of the machinery and equipment also guarantees its optimal efficiency,

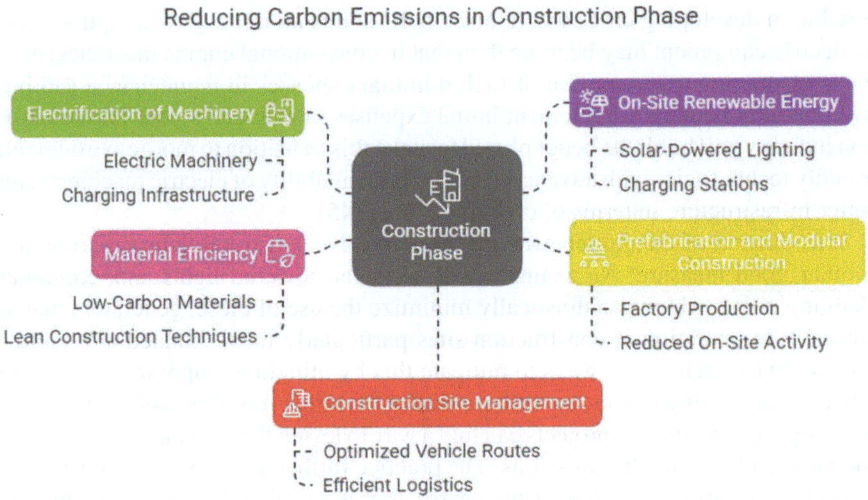

Fig. 5 Reducing carbon emissions in construction phase

decreasing the fuel consumption and emission. Through a whole society approach to the construction process, the industry will go a long way in diminishing its negative effect on the environment and be much more sustainable in future constructions.

2.5 Operational Phase

Operational period in the life cycle of a building is one of the main sources of carbon emission mainly through persistent energy demand to heat, ventilate and air condition (HVAC) system, lighting, water heating and turn-on of electric devices. Commercial, institutional, and large residential buildings, in particular, use large quantities of energy in their HVAC systems since it is necessary to provide thermal comfort to the building occupants at all times. Poorly designed systems, a lack of proper maintenance, and inefficient use of such systems cause redundant energy deposits and an increase in carbon emissions. On the same note, lighting and water heating, which are not controlled or optimized, add additional carbon footprint to the operational activity of a building [9].

In order to subside these tensions the use of smart building technology has risen as a vital way of countering it. This can be done by means of the Internet of Things (IoT) devices and sensors able to monitor and optimize energy consumption of the numerous systems deployed inside a building in real-time. An example of this is occupancy sensors which would determine that spaces are unoccupied and turn off the lighting and HVAC systems to conserve energy where it is not being used. Smart thermostats have access to control temperatures freely depending on occupancy and weather conditions, thus energy is never wasted. Such systems are flexible to alterations in the operating patterns, identify areas where efficiency needs to be enhanced and give workable data to the facility managers so as to bolster their operations. The utilization of these technologies allows the building to be run as efficiently as possible, and the energy consumption to fit the real demands instead of a preset schedule.

Green building certification, including LEED (Leadership in Energy and Environmental Design), BREEAM (Building Research Establishment Environment Assessment Method) and GRIHA (Green Rating for Integrated Habitat Assessment), is another key feature of lowering operational carbon footprint. These certifications offer rigidity in terms of how developers and building operatives manage to unleash energy saving initiatives, the use of eco-friendly building materials and incorporation of the best practices in terms of operational efficiency. These certifications do more than just validate how well a building performs on environmental aspects since they actively promote continued performance improvements in regard to energy management. Such buildings which qualify to acquire such certifications do register a long-term decline in energy used, and as a result, there is a decline in operational carbon emissions as time goes by [15].

Other than technology and certification, the integration of renewable energy is essential in the reduction of operational carbon footprint. Adding solar cells, wind power or geothermal heat and cooling system as part of the building design can go a long way to countering the energy requirements that are usually supplied using fossil fuel. Other than mitigating dependence on non-renewable resources, these systems will also aid in reduction of overall carbon emissions and this is in keeping with the global sustainability agenda.

Finally, the design of energy-efficient buildings is an essential part of decreasing the emissions of the operation. This entails high-performance insulation, energy efficient and window treatments, complex glazing systems, air tight structures which decrease heat loss during winter and cooling required in summer. Passive elements like passive lighting and passive-solar heating, also work to decrease the use of artificial lights and heating and cooling systems. Using such strategies the buildings can be kept comfortably with least energy consumed, thus minimizing the carbon footprint of buildings.

Overall, building operation is the potentially crucial point of carbon emissions reduction. With incorporation of smart technologies, renewable energy resources and energy efficient building techniques the running phase can go on to become a key component of environmental performance and sustainability as opposed to being a significant source of carbon emissions. With energy efficiency and the continuous improvement, buildings can largely reduce their carbon footprint, which leads to making the built environment more sustainable. Cycle of sustainable building operations is observed in Fig. 6.

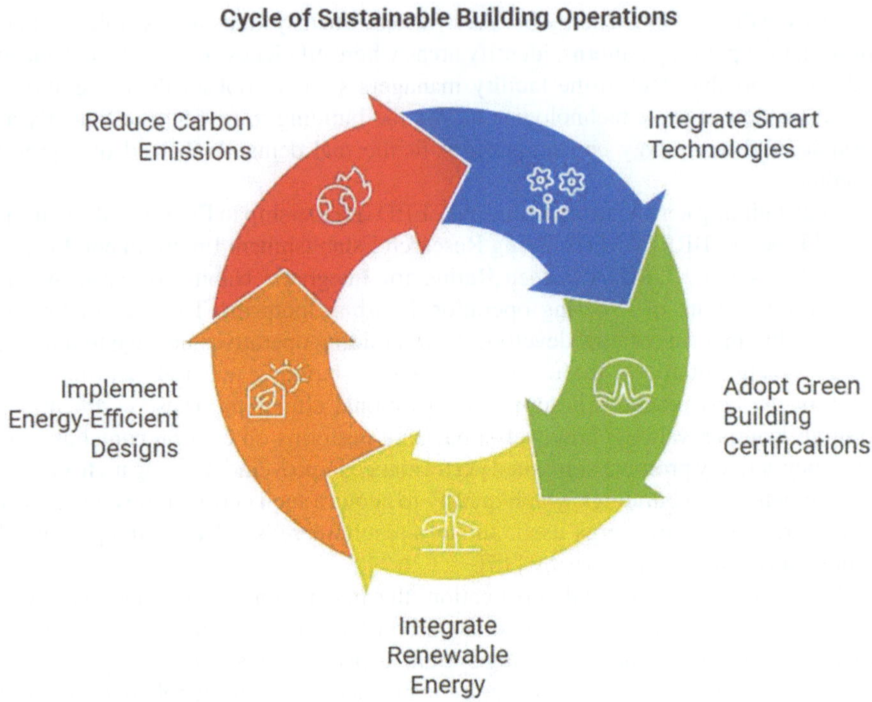

Fig. 6 Cycle of sustainable building operations

2.6 Maintenance and Retrofitting

The process of maintenance and retrofitting plays the key role in reducing building carbon footprints across whole life expectancy. Preventative maintenance helps in keeping all the systems installed in the building in good shape, particularly machines using much energy such as the HVAC (heating, ventilation, and air conditioning) systems, lights, and water heaters. The poorly maintained systems consume more energy not only had they tend to get shorter life but need to be replaced more frequently throughout their existence thus causing more consumption of resources and increasing emissions in the long-run. Thus, through implementing routine inspection intervals, conducting periodical servicing and practicing preventive maintenance, the owners of the buildings will be able to decrease the unnecessary energy consumption and increase the years of operation of the most crucial building systems. This has the benefit of minimizing the energy-demanding repairs and replacements that need to be done which promotes sustainability in the long term [32].

In addition to regular maintenance, retrofitting is crucial in decreasing carbon emission in the built environment, especially among the existing buildings which were initially developed without focusing on energy use. Most of the pre-existing structures that were constructed decades ago, in times when knowledge of green building was not in the mainstream, tend to harbor poorly efficient systems and materials that have long been outdated. The retrofitting of such buildings will involve provision of thermal insulation to minimise energy requirements in heating and cooling the building. A few of the more common retrofit techniques are air sealing, adding more wall and roof insulation, and installing high-performance glazing on old windows. All these variations aid in curbing loss of heat during winter and excessive and unwanted heat gain during summer and therefore greatly cut down energy use in heating and cooling [7].

The other critical component with retrofitting lies in the fact that building systems that are old and inefficient are replaced by modern and energy efficient systems. This may encompass retrofitting of old fashioned boilers, air-conditioning systems, lighting, power and electrical devices. An example is the replacement of old incandescent lighting with modern energy efficient LED lights or the replacement of old HVAC systems with inverter powered systems as these interventions would provide a lot of energy savings. Such efficiencies do not only decrease energy use but also assist in decreasing operational CO_2 emissions that comply with wider scaling of sustainabilities [37].

There is also a prospect to install smart technologies that allow a building to perform automatically and optimally with retrofitting. Occupancy sensors and programmable thermostats can also be used as well as highly sophisticated building management systems (BMS) that can increase efficiency in the control of energy consumed. Self-adapting heating, cooling and lighting systems can adapt to real time occupancy or environmental variables so energy is only used when and where it is required and in the most efficient manner. These smart solutions not only play the role of reducing the energy consumption but also enhance the comfort level of occupants by providing the best indoor environment and using minimal amounts of energy [30].

Retrofitting project when implemented carefully can greatly make a poor-performing building an asset with high-efficiency. Such a process impacts the environmental performance of a given building and helps ensure the decarbonization objectives of the cities and nations. In many cases, retrofitting is perceived as a cheaper and less disruptive way of demolition and reconstruction, and retrofitting is also a key to enhancing sustainability of the old building stock. Retrofitting contributes to the reduction of long-term sustainability goals through better energy performance and carbon emissions of the operations, a significant step towards fighting climate change. Maintenance versus retrofitting can be compared is compared in Fig. 7.

To sum the matter up, maintenance and retrofitting of buildings to help cut down the carbon footprint of building operations can hardly be over-emphasized. These measures also add to short term savings in energy use and CO_2 emissions as well as longer term savings by increasing the building life of systems and better overall building performance. Maintenance and retrofitting should become priorities of sustainability plans that would allow setting ambitious carbon reductions goals, which would be involved in the construction and building operation field.

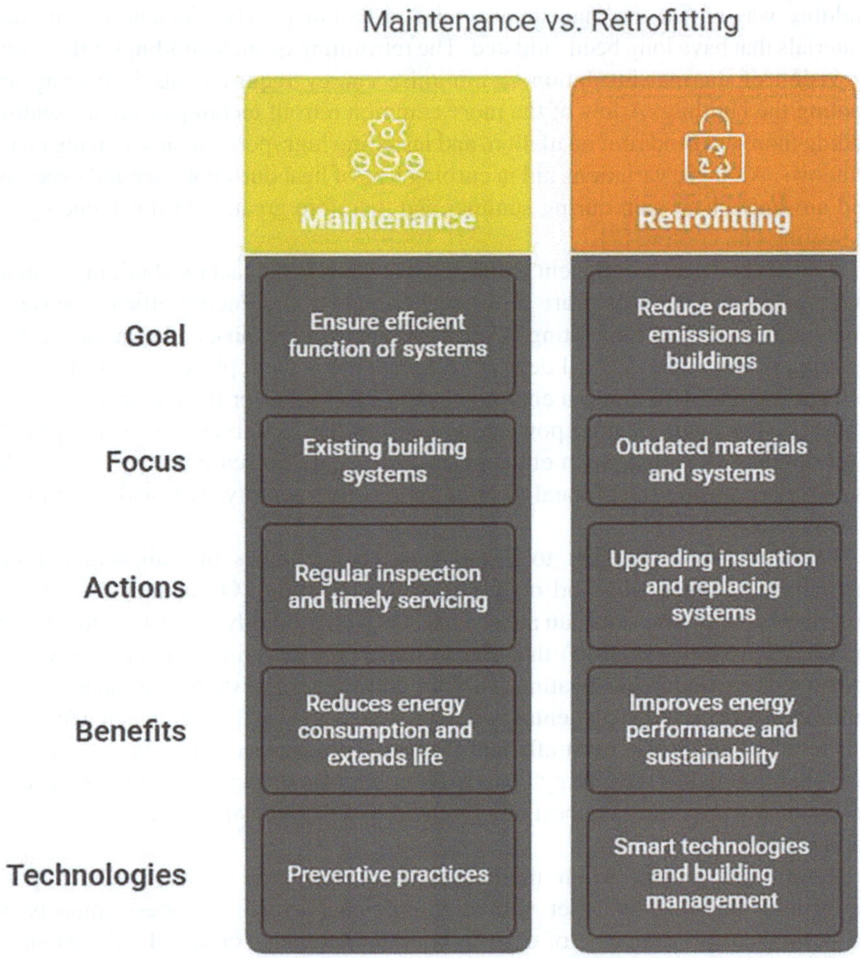

Fig. 7 Maintenance versus retrofitting

2.7 *End-of-Life (Demolition and Recycling)*

The demolition phase in buildings (including the re-use of materials and subsequent disposal), whereby a building is demolished and the materials cleared into the environment, is a vital aspect in the life cycle of the constructive industry but tends to be dropped in conversation regarding carbon reduction. With buildings near the end of their life cycle, razing them down entails heavy-duty equipment like excavators and wrecking balls that use high quantities of diesel fuel and release greenhouse gases. This occupies huge amount of construction and demolition (C&D) waste comprising of concrete, brick, steel, timber, plasterboard, glass and insulation materials. A lot of this wastes end up in landfills where the decay process may result in putting stored or embodied carbon back into the atmosphere. Also there is waste transportation, storage and disposal logistics that contributes to the burden on the environment even further [4].

In order to reduce the carbon energy footprint of demolition, most construction stakeholders now promote strategic alternatives like deconstruction. Deconstruction, in contrast to conventional demolition, is more labor-intensive and controlled form of deconstructing building by carefully taking them apart in an attempt to salvage the integrity of the materials, so that they can be reused. The pieces of construction materials like timber beams, bricks, steel items, windows, and doors, can be saved and reused either in other construction works or in renovation works. Not only does it prolong the death of precious materials but also helps in curbing the usage of precious materials since there is a fall in the demand of virgin material as well which results in the carbon footprint, created by the production of new materials, being avoided. Dust and particulate emissions are also reduced on the site of deconstruction that enhances environmental and health conditions of workers [12].

The other aspect that is important towards cutting end-of-life carbon consequences is recycling construction materials. Crushed concrete example can be used as recycled aggregate in newer concrete mixes or even used as road base material avoiding the need to quarrying fresh supplies. Steel, having unlimited strength after recycling is available and can be re-melted and used again in buildings. Properly treated timber, especially with preservation, can enter long-term second life as furniture, cladding, non-load bearing structural members. These recycling activities go a long way towards helping in the circle economy where materials are reused in the same value chain rather than being treated as waste [5].

Design-stage inclusion of sustainability can also make the environmentally friendly end-of-life management much easier. Design for Disassembly (DfD) is a new way of thinking in design that seeks to build buildings that can be disassembled with a proper modularity, reversibility and the ease to be decomposed. Such measures include the adoption of mechanical fasteners instead of permanent glues, the marking of building parts that are easily recognizable as well as the design of connections that can be reversible with the use of connecting elements without materials being damaged. Structures designed with the concept of DfD can not only enable the efficient separation of materials at the end of life cycle but also can enable the more successful adaptive reuse and design flexibility in the future (changes or repair) [35].

Furthermore, numerous progressive cities and privately financed developers are starting to reward reuse and recycling by means of policies, tax credits and recycling diversion goals. Stricter regulations on the use of the landfills, the cultivation of the schemes of the certification of building materials, and the investment in the recycling infrastructure will help make the industry much greener in terms of its end-of-life processes. It is also important to educate and train demolition contractors, architects and developers to undertake sustainable demolition activities on a large scale.

To sum up, the final stage of the life of the buildings is both challenge and opportunity to reduce carbon in the construction sector. By leaning forward into proactive solutions like deconstruction, recycling of materials, and design versus disassembly, meaningful achievements can be attained that would help in minimalizing

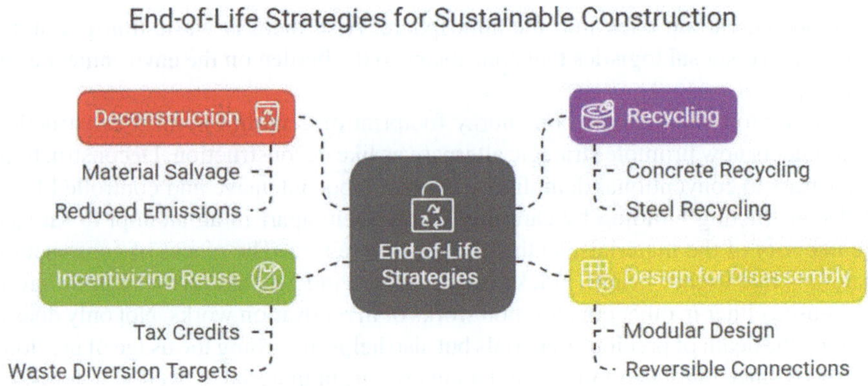

Fig. 8 End-of-life strategies for sustainable construction

waste, saving resources, and cutting emissions. The practices are not only considered towards climate action, but they also fit within larger sustainability, resilience, and resource efficiency, in the built environment. End-of-life strategies for sustainable construction is observed in Fig. 8.

2.8 Carbon Footprint in the Construction Lifecycle

Sustainability of buildings has recently taken a vital direction especially on the carbon footprint of the building throughout the full life cycle. Construction industry is one such sector which contributes to a large extent to environmental degradation and is estimated to cause 30–40% of emission of green gas globally. The mechanisms and processes of releasing carbon emissions in the lifecycle of buildings as well as providing answers on how emissions can be optimized or minimized are very important to the development of sustainable construction [31].

Structural system is one of the major factors that determine the life cycle carbon footprint of the building. Some of the main features (selection of materials, type of lateral load-resisting system, the height of a building) affect directly carbon emissions. They are normally arrived at early during the design phase and influenced by various factors such as structural efficiency, cost, construction schedules and design needs. Nevertheless, even at the time when the concept of sustainability becomes increasingly popular, concerns regarding the environment such as carbon emissions are not systematically incorporated into the process of selecting the structural system. This difference is particularly significant given that building carbon (the amount of carbon emitted through material extraction, processing, transportation, and the resulting construction) and building operational carbon (carbon emitted through the operation of a building with regard to energy use over the lifetime of the structure) is a huge factor in a building overall carbon footprint [33].

An important role is played by embodied carbon, which is frequently not taken into account. Especially during the early and late phases of the whole lifecycle. It includes emissions made throughout the production and construction process of the materials it uses, and once more throughout demolition and waste disposal. Operational carbon in contrast is more dominant in the use stage and has over the years been given more priority because of the nature which is long term and less complicated to measure. Nevertheless, the two forms are necessary in comprehending the total environmental cost of building.

There is an increased international and regional action to minimize carbon emissions due to the built environment. As a case in point, the European Union has ambitious targets like achieving 40% reduction of greenhouse gases by 2030, greater utilization of renewable energy and a massive enhancement of energy efficiency. In this effort, almost zero energy buildings (nZEB) have been a key item. These constructions aim at reducing the needs in operational energy by offering energy efficient design and integration of renewable energy. New ideas such as zero-carbon buildings take this further and add life cycle carbon evaluation to minimize embodied and operational emission [48].

Sustainable design has still some issues even though there are improvements in the field. A significant source of emissions, which is ignored in carbon accounting analysis, is the construction delivery stage where intensive use of machines, transportation and temporary structures is implemented. This is typically a highly concentrated, large emission source and there are large mitigation opportunities to be realised through use of better logistics, material utilization, and low-carbon technology.

The other vital dimension is management of construction waste. Some sources of wastes include construction, renovation, and demolition wastes that may result in air and land pollution besides wasting valuable resources. The backlash on the environment is eliminated through effective construction waste management actions, including reusing, recycling and effective disposal. Recycling of the concrete and reuse of the bricks and careful planning of construction can greatly improve the embodied carbon as there is less dependence on virgin materials and processes that are energy intensive.

Moreover, a life cycle comparison of carbon footprint generated by building new constructions and refurbishing existing constructions will disclose the notion that refurbishment of existing buildings is sometimes more environmentally friendly. Much of the new construction stock to come in 2050 is already present today in many countries including the United Kingdom. Thus, to meet climate objectives, it may be necessary to shift the emphasis away from the construction of new, energy-efficient buildings and direct more attention toward the huge Pool of existing buildings. Life cycle assessments indicate that time and again costs of refurbishing and upgrading current buildings have less carbon footprint than that of demolition and the construction of newer buildings that are made to be more energy efficient [39].

A cradle-to-cradle approach is necessary to really address the carbon footprint of buildings throughout their lifecycle (including thinking about life cycle considerations when making initial decisions through construction, use, and ultimate deconstruction). Environmental performance should be added as a critical design determinant in structures. It is necessary to change construction practices turning to low-carbon building materials and avoiding excessive wastes. Lastly, policy frameworks should start looking beyond the operational emissions to cover the embodied effects in full.

To conclude, the goal of realizing a low-carbon construction industry entails an environmental system transformation. Whether it is the selection of an overall structural system or the level of waste management or refurbishment strategies, each stage of a buildings lifecycle throws up challenges and opportunities to reduce carbon. Based on proper data and new technologies combined with effective regulatory structures, a life cycle approach can inform the industry on more resilient and climate-sustainable operation.

The effects of construction industry to the environment are quite significant, especially when considering green house gas emissions. With increasing urbanization occurring all over the world, buildings and infrastructure are increasingly using a lot of resources and energy. The construction industry is said to contribute to 30–40% of carbon dioxide emissions in the world and therefore it presents an area of concern when it comes to climate change. The carbon footprint must, thus, be addressed in the life cycle of a building in order to bring about environmental sustainability and also to achieve the global emissions goals.

2.8.1 Structural System Design and Its Carbon Impact

The initial decision made which is one of the most heavily felt choices in the development of a building is the choice of structural system. The type of material, e.g. steel, concrete or timber, as well as the lateral load-bearing mechanism and the height of the building, play a role in its total life cycle carbon emissions. Such decisions are usually prompted by performance, safety, cost, speed of construction, architectural objectives. Nevertheless, issues of the environment are seldom at the heart of the decision since the focus is mostly on the carbon footprint of these systems.

Though concrete and steel reign supreme in the industry, they are one of the most carbon-intensive materials. On its own, cement manufacture makes up roughly 8% of worldwide emissions of CO_2. By contrast, new materials including cross-laminated timber (CLT), geopolymer concrete, and recycled steel all present opportunities to embark on very large reductions in embodied carbon. Designers are now being asked to find a balance between more conventional measures of structural performance and carbon implications by selecting systems that are both economically and ecologically sound [42].

2.8.2 Stages of Life Cycle of Carbon Emissions

Carbon emissions relating to buildings are usually grouped into two broad categories:

- Embodied Carbon: Emissions deriving out of the extraction, production, transportation, construction, maintenance, and end-of-life processing of materials. They can be front-loaded into the life of a building, but constitute a large part of overall emissions.
- Operational Carbon: Carbon content of energy use at the occupation stage of the construction such as heating facilities, chilling, lighting, fittings, etc. Analyzing historically, they have been accorded maximum attention since they have contributed in the long term, and they directly relate to energy consumption.

A genuinely sustainable building ought to compensate the two. Such tools as Life Cycle Assessment (LCA) can be used to measure and compare emissions at each stage and make evidence-based decisions. In other words, an embodied carbon reduction so far into early stages can result in cumulatively impactful results, especially in high-rise or infrastructure projects, which demand large amounts of material [22].

2.8.3 Policy/Regulation Role

The policy interventions have been instrumental in enhancing the construction of low-carbon constructions. The "2030 Climate and Energy Framework" put forward by the European Union works towards a 40% decrease in greenhouse gas emissions compared to the year 1990, and progress to greater renewable energy consumption and energy efficiency. Building performance regulations like Energy Performance of Buildings Directive (EPBD) demand that all new buildings have an approach that will enable them to reach nearly zero-energy targets.

The administration of numerous countries has initiated carbon accounting in the building and contract policies. LEED, BREEAM, and EDGE are also voluntary certifications that motivate the construction industry to incorporate more green building construction and award points on low impact materials, carbon efficient design and integrating renewable energy.

Nevertheless, the orientation of policy is still usually biased to operational rhetoric of energy performance. There would be a broader change required to impose the limit of embodied carbon, particularly to large-scale construction. New laws in certain jurisdictions even now require whole-life carbon reporting which is an essential step to achieve carbon management.

2.8.4 Emissions During Construction Phase and Reductive Prospects

Construction delivery phase is rather one of the carbon-intensive phases that include associated fuel, large equipment, logistics, and complicated systems. The duration in this case is short, but carbon emissions can be underestimated, and they are quite substantial because of the intensity and the concentration of activity. The electrification of construction sites is becoming more and more popular, as well as the use of low-carbon concrete and modular or prefab construction as a method of minimizing emissions at this stage.

Digital technologies like Building Information Modeling (BIM), Digital Twins, and Integrated Project Delivery (IPD) now allow us to accurately simulate, monitor and streamline carbon emissions even prior to a foundation being poured. These tools assist in ensuring material efficiencies and minimizing wastes as well as improving coordination among the stakeholders with less effect on the environment both during the lifecycle of the project.

2.8.5 Being Deprived of the Promise of Refurbishment

Although it is often admitted that the focus of attention should be on the construction of new buildings that will be energy-efficient, in most situations refurbishment may provide more returns on investment in terms of sustainability. Putting modern insulation, windows, mechanical systems and renewable energy sources in older buildings can save a staggering quantum of operational carbon and avoid the embodied carbon of demolition and rebuilding.

The existence of a high percentage of the existing building stock especially in the United Kingdom necessitates refurbishment measures. Literature indicates that three out of every four buildings that are anticipated to be in existence in 2050 are found today. When well taken care of, refurbishment may help increase the useful life of buildings, save on urban wastes, and make significant contributions on carbon reduction targets.

2.8.6 Inclusion Within a Construction Waste and Circular Economy

The activities that result in the construction and demolition represent a large source of the global solid waste. Ineffective waste management does not only contribute to the volume of wastes in landfills, but also causes the ineffective transport and the degradation of materials, which emit unnecessary gases. The carbon footprint of a construction project can be substantially minimized through a circular economy approach, i.e., when the resources of the construction site are prioritized to adopt a resource recovery, reuse, and recycling measure.

On-site sorting, sorting concrete and metals and recycling, repurposing salvaged materials, design for disassembling are not only major components of green construction but also come in handy during construction activities. These strategies also contribute to the preservation of natural resources and lowering of the reliance on the extraction of virgin materials which, in most cases, is energy-demanding and environmentally disrupting.

2.8.7 The Role of Stakeholders to Carbon Mitigation

To achieve winning results over carbon footprint in the construction lifecycle, it is necessary that a concerted action by everyone who is a stakeholder in the process is sought. All of the above play a vital role in collaborative project delivery, shared responsibility responsibilities and open books in carbon accounting.

The clients need to be informed to consider life cycle sustainability rather than initial cost. Tools and methodologies that design to carbon model should be adopted by designers. Contractors and builders are recommended to invest in low-carbon and operate more efficient sites. It is necessary to persistently reinforce policy and financial incentives to support sustainable practices by the government and municipalities.

3 Cross-Cutting Strategies for Emission Reduction

3.1 Life Cycle Assessment (LCA)

Embodied carbon is the sum of greenhouse gases that are required to produce and deliver building materials-including extracting raw materials to the manufacturing and transport process as well as the on-site assembly. Embodied carbon must be built in before someone moves into the building unlike operational carbon that would be emitted once the building has been occupied. It is a large part of the life cycle carbon footprint of a building, and a significant fraction in energy efficient modern buildings where operational emissions may be a small fraction. It is relevant to understand and treat embodied carbon at the concept stages of the project planning and design because most of the decisions that impact on such carbon are determined before the construction commences [8].

The construction sector can be said to be the largest contributor of the embodied carbon due to the high emission substances like cement, steel and aluminium. An example is the manufacture of Portland cement which is the main component of concrete, thus consuming fossil fuel during its process besides emission of CO_2 during activation of limestone, which makes it one of the most significant producers of carbon in the industrial world. Equally, steel and aluminium manufacturing consume large amounts of energy, which usually come in the form of non-renewable energy sources. In comparison, building materials made of timber, bamboo, or recycled content equivalents can generate much less embodied carbon and still achieve decent performance when sustainably produced.

Embodied carbon reduction should be approached strategically; in this regard, it is possible to mention material optimization, reuse of existing buildings or parts, local supply, and the integration of low-carbon technologies. Also, design choices that minimize the total amount of materials and resources consumed, via achieving efficient structural systems, modular construction and lightweight options are crucial.

Moreover, applying any new technology like carbon-capturing concrete, geopolymer binders, and new types of insulating materials, one can significantly decrease the amount of embedded emissions of new constructions.

As outlined in the prior section, Life Cycle Assessment (LCA) is a crucial process of embodied carbon measurement and pinpointing of the most significant links. With the help of LCA, design teams will be able to contrast possible materials and construction methods with regard to environmental performance and make decisions based on the information obtained. Beyond these, Environmental Product Declarations (EPDs) provide standardised information on the embodied impacts of a particular product and can assist an architect or engineer to make sustainable decisions at the specification level use.

Embodied carbon metrics are appearing in the frameworks of governments, green building certification programs and sustainability rating systems. LEED v4, BREEAM and the Indian GRIHA system standards promote or require reporting and minimisation of embodied carbon. Others are even progressing toward capped embodied carbon regulations of particular building types or, building-permit documentation requirements. Such dynamic demands make the industry move towards greater responsibility of sourcing, transparency of manufacturing, and construction materials innovation [10].

In conclusion, embodied carbon reduction can be considered one of the main elements of sustainable construction that can support the process of enhancing the operational efficiency. Project teams can greatly minimize building and infrastructure impacts over the longterm by acting mindfully early in the design and the procurement process. This change would not only help achieve the desired climate goals globally but also mean that the construction sector would be taking a front row role in terms of proper environmental stewardship. Reducing embodied carbon in construction is mentioned in Fig. 9.

3.2 Carbon Capture and Storage (CCS)

CCS (Carbon Capture and Storage) is quickly developing into a game changer in the quest to make the construction industry more environmentally friendly. With a heritage in fossil-based power generation and heavy industry, CCS is emerging in purpose-built environment where it is considered to offset embodied and operational carbon pollution. The strategy implies seizing the production of carbon dioxide, either at the construction stage of materials or during the functioning of a building itself and storing it or utilizing it long-term. Since environmental responsibility has become an essential consideration in construction practices, CCS technologies are also an encouraging avenue not just towards diminished production of emissions but towards a possibility of transforming buildings into carbon-sinks.

The creation of the carbon-absorbing concrete hits the list of the most prominent achievements in this field, e.g., that which can be seen in the technologies, like CarbonCure. In this process, CO_2 gas that is captured is added to concrete as it

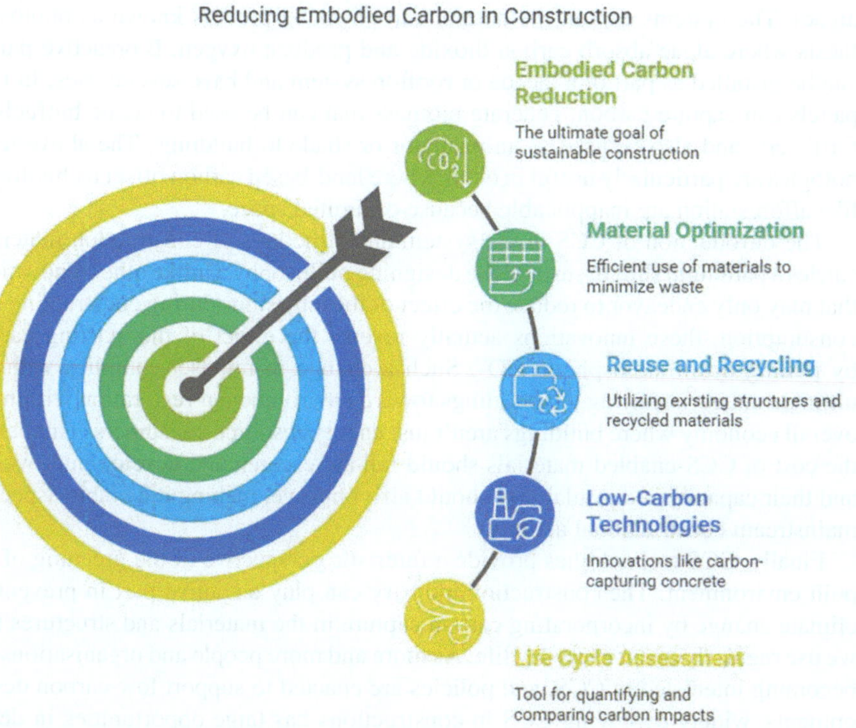

Fig. 9 Reducing embodied carbon in construction

is being mixed and reacts with calcium ions to form calcium carbonate—a stable compound permanently entrapped in concrete. This not only sequesters carbon, but also enhances the material with compressive strength, which means the material requires less cement. Since cement production is among the single greatest source of embodied carbon in constructions, this innovation has the advantage of cutting down the results of material processes, besides improving on the structural performance [41].

In addition to concrete, the use of CO_2—absorbing materials in architectural elements is picking up steam. Some of the façade systems are even designed in such a manner using mineral based compounds that chemically bind to airborne CO_2 and over time convert them to inert forms. This front represents passive carbon capture as the front removes small quantities of carbon dioxide constantly out of the surrounding air. Certain materials go one step further, having photocatalitic coatings added, to harness sunlight to speed-up these reactions. Such systems placed on the exteriors of buildings do not only enhance environmental performance but have an added benefit of a higher level of air quality within a highly populated area.

An especially novel subcategory of CCS is what has been referred to as bioreactive systems: building materials or panels impregnated with live organisms (typically algae). The systems exploit the natural carbon fixing process known as photosynthesis where algae absorb carbon dioxide and produce oxygen. Bioreactive panels can be installed as part of a façade or rooftop system and have several uses: In fact, panels can capture carbon, generate biomass that can be used to create biofuels or fertilizers, and also be used as an insulator or shade to buildings. The above technologies are particularly useful in cities where land-based carbon offset technologies like afforestation are inapplicable because of limited space.

The introduction of CCS in the system and structures of construction materials carries a paradigm shift in sustainable designing philosophy. Unlike other innovations that may only endeavor to reduce the effect of the emissions during constructing and consumption, these innovations actually reverse the effect of the existing carobs by pulling down atmospheric CO_2. Such a change in mindset complies with the ultimate net-zero shifting the buildings toward environmental restoration within the overall economy where buildings aren't just energy/resource consumers. Thereafter, the cost of CCS-enabled materials should fall as research and development works, and their capacity and availability should also improve, making it a viable aspect of mainstream constructional activity.

Finally, CCS technologies provide a futuristic perspective of the greening of the built environment. The construction industry can play an active part in preventing climate change by incorporating carbon capture in the materials and structures that we use regularly in our everyday life. As more and more people and organisations are becoming interested in CCS, and policies are enacted to support low-carbon developments, widely deploying CCS in constructions has huge opportunities in delivering climate resilient and future-proof infrastructure solutions. CCS technologies in construction is mentioned in Fig. 10.

3.3 Water-Energy Nexus

Water energy nexus is a dynamic and mutually dependent interrelation between energy consumption and water consumption as both are inherent to the sustainability of the construction industry and building operations. Water provides several uses in the built environment besides daily domestic and commercial activities since a high amount of energy is needed to extract, purify, distribute, heat and treat wastewaters. On the other hand, most energy facilities are highly dependent on water to cool, process and wash especially within the electricity generation and the industrial sectors. Most of the time, lack of efficiency in one area tends to result in wastage of resources in the other. The identification and maximization of such a linked system plays an important role in minimizing the collective environmental impact and the attainment of resource-efficient building performance [34].

The introduction of the advanced water-saving technology is one of the most effective strategies that can be used to increase the efficiency associated with water-energy nexus. This comprises greywater recycling systems being adopted where waste water is gathered on relatively clean sources like laundry machines, showers

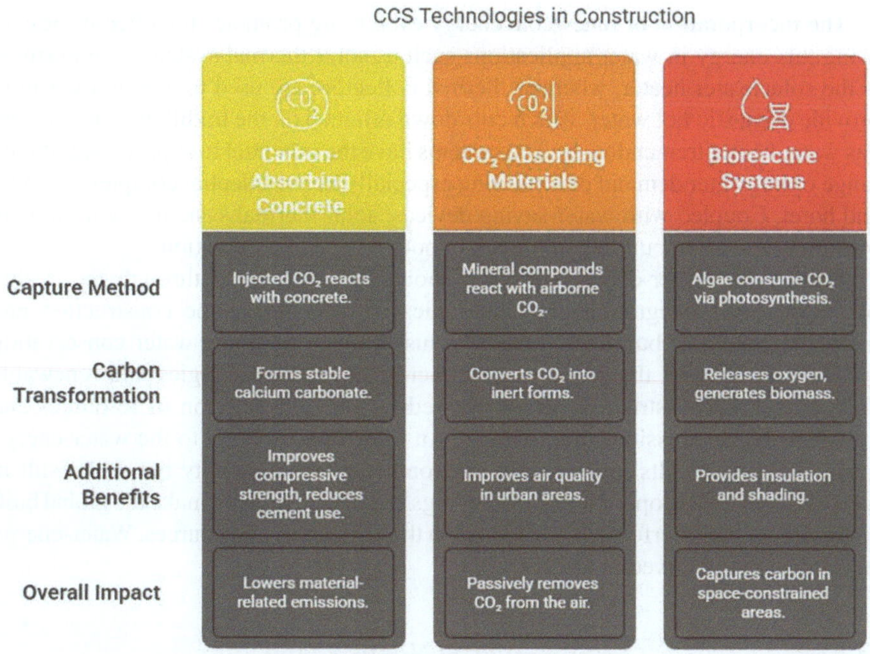

Fig. 10 CCS technologies in construction

and bathroom sinks. After being treated, such greywater can be used in non-drinkable ways such as in toilet flushing, irrigation or even refrigeration systems in commercial structures. Greywater systems also greatly reduce the energy footprint of municipal water supply and sewage treatment, by reducing the amount of fresh potable water needed, in addition to the amount of waste water to treatment plants.

At the same time, low-flow plumbing fixtures are crucial in reducing the use of water as well as the consumption of energy. Low-flow fixtures like faucets, aerated showerheads and dual-flush toilets are designed to provide the same performance at significantly lower flow rates. These devices save not only on water consumption but on energy needed to heat up water, in particular hospitality and residential buildings. The net result is that friction is obviously less in terms of utility bills and a lesser total carbon footprint, converting these fixtures into a money- and earth-saving improvement.

In addition to internal systems, there is another sustainable approach also directly aiding the nexus of water and energy, that is, rainwater harvesting. Through collecting and storing rainwater in its systems, which can be used in cooling systems, toilet flushing and landscape irrigation, buildings can stop depending on the municipal water supply thus decreasing dependence. This, subsequently, relieves load on the centralized infrastructure and the energy cost to pump and treat massive amounts of water. Rainwater systems can be particular advantageous in the areas where water shortage occurs or where energy costs are very high to transport water.

The incorporation of renewable energy even more promotes the effectiveness of using this energy in water applications such as solar thermal systems. An example is the solar water heater, whereby thermal collectors are used to exploit the sun to provide domestic hot water, which cuts down reliance on the traditional electric and gas water heater tremendously. The systems have the potential to supply a significant range of hot water demand of a building especially in a residential complex, hospital and hotel. Coupled with water-saving devices, solar thermal systems are used in an overall, low-carbon building approach in both design and operation.

Finally, the water-energy interface should be envisioned through the design of systems and integration of technologies so as to make the construction and the management of buildings genuinely sustainable. Applying water conservation methods along with the promotion of energy-saving technologies and renewable energy use, the construction sector can reduce the consumption of resources and greenhouse gas emissions dramatically. An integrated strategy to the water-energy nexus not only results in long-term environmental sustainability but will result in increased resilience, operational cost savings and aids in helping make the global built environment high-performance, efficient in the utilization of resources. Water-energy nexus cycle is observed in Fig. 11.

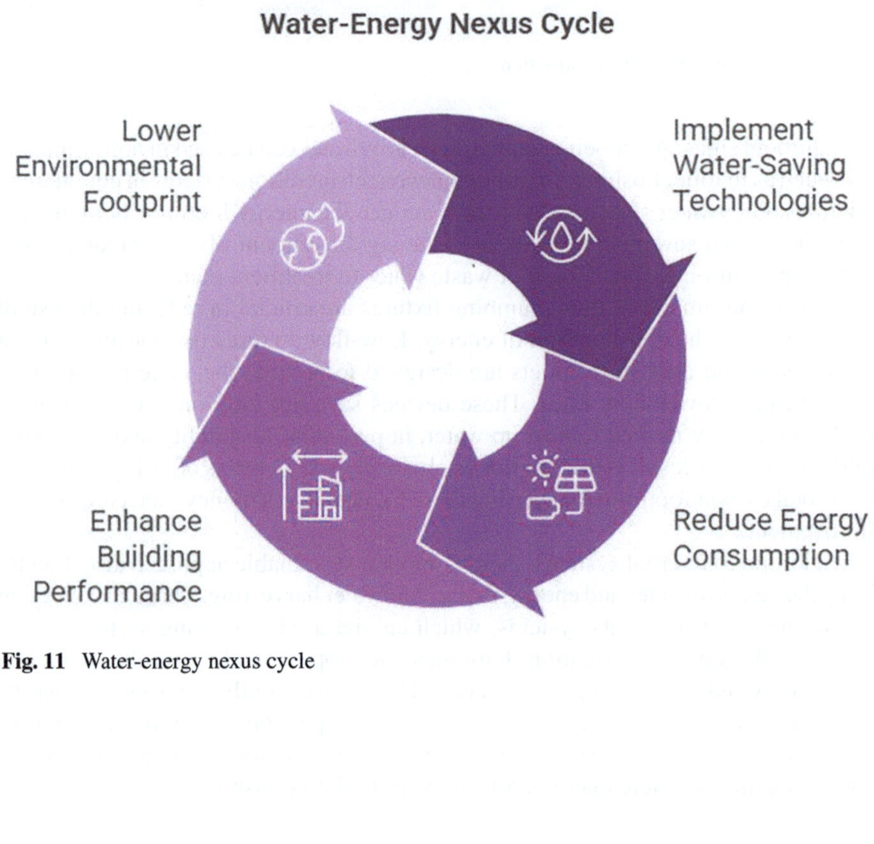

Fig. 11 Water-energy nexus cycle

3.4 Waste Management

The efficient handling of waste is crucial in curbing the environmental cost of the construction industry, especially in countering the excess amount of carbon emissions and dealing with the enormous amounts of construction and demolition (C&D) waste. C&D waste constitutes a large percentage of the total solid waste in the world, and in the absence of adequate management, goes to landfills where it will produce greenhouse gases via decomposition and generate virgin products, which demand increased resources. Enforcing effective waste management strategies not only brings about environmental objectives but also promotes effectiveness and economy of the project [18].

On-site segregation of C&D waste happens to be one of the most practical and effective methods. At the source, i.e. categorizing waste materials like concrete, bricks, metals, timber, plastics and drywalls construction sites can work towards ensuring that recyclable or reusable materials are well separated with the non-recyclable one. The practice makes the process of recycling easy and cuts drastically the amount of garbage that makes it to the landfill. It also enhances the value of the recovered resources and makes them more valuable to be utilized again in construction or industry by ensuring that they are of better quality thus reducing the embodied carbon incurred in the manufacturing of new materials.

The other important sustainability strategy employed in waste management is use of recycled aggregates especially crushed concrete and masonry. Upon processing, these materials may be good alternatives to natural aggregates in many construction activities, such as concrete manufacture, road Sub-base and foundation levels. The use of recycled aggregates also saves natural resources and energy expenditure and emissions of the virgin materials as well as quarrying and transportation of the material. As recycled materials can be gathered locally, the strategy also leads to reduction of transport-related emissions and therefore increases the carbon saving advantages.

Moreover, this applying circular economy principles to constructions is picking pace. A circular economy deals with prolonging the utility and life of materials by employing methods like reuse, refurbishment and recycling. Partnering with circular economy networks enables the construction companies to participate in material exchange programs with which materials that are not used, excess, or reclaimed during a construction process are being diverted to other projects rather than being wasted. This is because this closed-loop system does not create much waste, encourages material recovery, and lengthens the life cycle of building resources greatly, which suit the goal of sustainable development.

Technology has extended its role towards waste management as well. Real-time monitoring of the generation and disposal of waste can be done by the use of digital tracking systems, material flow analysis tools, and waste management apps. The tools assist construction managers in making more satisfactory decisions and tracking down their waste reduction goals and whether they are complying with environmental regulations or not. Leveraging data allows corporations to determine primary

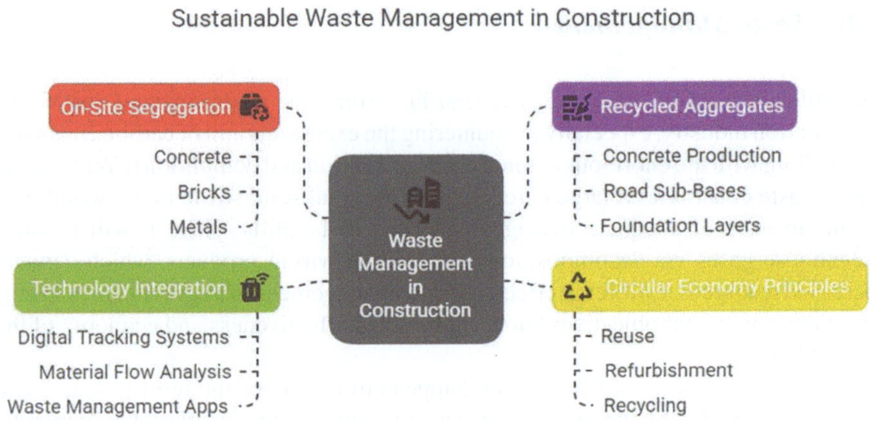

Fig. 12 Sustainable waste management in construction

waste streams, use resources more efficiently, and perpetually enhance the level of sustainability.

To sum up, the issue of waste management in the construction industry is not limited to simply removing it, but it also forms part of sustainable building. Implementing on-site segregation, utilization of recycled materials, achieving circular economy systems, and technological improvements, the industry will have a chance to minimize its carbon footprint and save substantial amounts of valuable resources. The broad application of these strategies can improve environmental conditions as well as make the construction industry more resilient and resource-efficient in general. Sustainable waste management in construction is mentioned in Fig. 12.

3.5 Carbon Footprint in the Construction Lifecycle

3.5.1 Limestone Quarrying: Carbon Footprint

Limestone is yet another type of rock that produces a lot of products being extracted which include the aggregates, lime, cement, and even construction blocks. Carbon emissions are mainly related to limestone mining because of the amount of diesel and electricity consumed during the disintegration, haulage, and processing of excavation apparatuses. The processes in the quarry, where the energy intense equipment (loaders, crushers, and haulers) uses extensive fossil fuel resources, is indicative of energy-intensive work.

In-depth life cycle analysis of limestone quarrying demonstrated that transportation processes are the main sources of resource depletion and generation of greenhouse gases (GHG) at the limestone quarrying. The total output had been measured into around 3.13 kg CO air equivalent per ton of crushed limestone. The conversion to the renewable energies including solar, biodiesel, or thermal could, potentially, cut the emissions by 0.21 million tons of CO_2/year [26].

3.5.2 Carbon Footprint of Cement and Concrete Manufacturing

Cement production contributes to high CO_2 emissions with about 5% of the world carbon dioxide being emitted due to cement production. Winning of Portland clinker (which is a major component of cement) emits nearly its weight in CO_2 represented mostly by the calcination process that contributes approximately 0.55 kg of CO_2 per kilogram of clinker.

A concrete, composite material made of sand, crushed stone, water and cement (around 12% by mass) material, has an average footprint of 347–351 kg CO_2-equivalent per cubic meter. Research has indicated a very high emission rate in terms of construction that include construction of multi-unit residential buildings and road constructions. As an illustration, in one instance, concrete of more than 312,000 kg of CO_2-equivalent gas was reported to have been emitted in the construction of residential houses, whereas in another case, 8215 metric tons of CO_2-equivalent gas had been released in the creation of every single kilometer of a Portland cement concrete pavement. Most emissions were found at the raw material extraction (92.7%), then in concrete production (7.2%) and a small portion of it was found in the construction at the site (0.1%).

3.5.3 Carbon Footprint of Asphalt Production and Construction

All the same, it is to be remembered that one can find one hundred percent, two hundred percent, three hundred percent, four hundred percent, five hundred percent, or even more than five hundred percent in a given financial maturity. Carbon Footprint of the asphalt production and construction.

The petroleum-derived material in road building, asphalt, is a large contributor to GHG emissions associated with construction. Manufacture of asphalt binders requires an amount of energy of approximately 4900 MJ/ton and 285 kg of CO_2 emissions/ton. Aggregates heating alone contribute 67% of such emissions, asphalt heating (14), and mixing (12).

On a 20-km highway construction project, emissions amounted to more than 52 million kilograms CO_2-equivalent. The biggest one was mixing operations, which added 54%, followed by 43% through raw materials production, and 3 more percent through the construction activities that include laying and compaction [29].

3.5.4 Steel Making Carbon Footprint

Another carbon-intensive process is steel production, largely as a result of the chemical reduction of iron ore in blast furnaces in a process that uses coking coal. This step alone has a proportion of 70–80% of the complete emission due to CO_2 emission in manufacturing steel. In the world around 6% of the total CO_2 emissions is attributed to the production of steel.

The carbon footprint can be greatly influenced by steel used in regular residential building. To illustrate this, use of steel B 500S in a medium size residential complex furnished almost 282,000 kg of CO_2-equivalent in a year. A closer examination of integrated and secondary steelmaking processes demonstrates that they require rather large amounts of energy and release an amount of emissions that is also quite substantial, which is why it is necessary to identify more sustainable options.

3.6 Mitigation Strategies in Construction Materials and Chemicals

3.6.1 Cement and Concrete Alternatives

The goal of reducing the emissions as a result of cement production is to pursue partial substitution of the clinker with industrial by-products, including fly ash, slag and silica fume. Desirable results have been exhibited by a replacement rate of up to 30% and the data do not take away strength. Another innovation is that alkali-activated slag (AAS) mortar drastically lowers fuel requirement and emissions as compared to conventional cement.

Oxy-fuel combustion systems are other technological interventions that could highly reduce energy requirements and emissions of nitrogen oxides. To illustrate, fluidized bed kilns save 10–15% in energy costs, and oxy-fuel saves up to 726 kg consequential to the ton of cement, although at extra cost of energy consumed in electricity.

Light concrete using waste products like old tires, crushed glass or demolition waste can emit up to 50% less emissions and improve mechanical properties and strength at the same time. Successful implementations of unconventional additives such as titanium dioxide, which potentially can diminish urban air pollution in the light of the photocatalytic process, have been presented in some studies [46].

3.6.2 The Modification of Asphalt

Additives such as Sasobit or synthetic zeolites make Warm Mix Asphalt (WMA) technologies, which lower the mixing temperatures and CO_2 emissions. WMA has even demonstrated an output of between 450 and 550 ppm of CO_2 as compared to the 700–750 ppm of traditional Hot Mix Asphalt (HMA). Fuel savings and a decrease in occupational hazards can also be translated to lower mixing temperatures.

3.6.3 The Construction Material Based on Clay

Brick, in particular fired clay bricks, is widely used on a huge energy expense in production. Such a sustainable alternative is unfired clay bricks which consist of raw clay soil and low-carbon binders. Magnesium oxide (MgO) is proving to be a good potential alternative to calcium-based binders because of the reduced calcination temperature and prominent heavy metal stabilization capability.

The CO_2 emissions can be minimized to almost 10 kg CO_2 per functional unit by using MgO based unfired bricks, as compared to the traditional bricks.

3.6.4 Recycled Aggregate Concrete (RAC)

The reduced use of virgin aggregates is offered by the RAC consisting of debris of concrete and other demolition products. It typically has worse properties, but by addition of pozzolanic admixtures, such as silica fume or fly ash, its behaviour can be enhanced. Researchers have indicated that emissions will be reduced by more than 22,000 tons per year within localized areas through the use RAC with appropriate additives [27].

Such innovations as Pixelcrete that uses up to 60% Portland cement with reclaimed materials exhibit dramatic carbon offsets up to 73.8 tons of CO_2-equivalent in one office building.

3.6.5 Asphalt Pavement Recycling (RAP)

RAP allows the use of aged asphalt in the deconstructed pavements. It may be reused made into new mixes or utilized as aggregates in base layers. Virgin asphalt has a carbon footprint of approximately 132 kg CO_2-equivalent to one ton of material, and containing RAP, the carbon footprint drops to about 84 kg per ton. Existing technologies in the minimization of emissions are further supplemented by streamlining aggregate sizes and applying cold recycling techniques.

3.7 Green Building: A Pathway to Low-Carbon Development

Dubbed sustainable buildings or high-performance buildings, green buildings are built to mitigate any harmful effect on the earth and the well-being of human beings. Such buildings are made in view of efficient energy, water and other resources consumption with minimized wastes and pollution as well as degradation of the environment along the lifecycle of the building. Whenever possible, green buildings tend to align with existing natural ecosystems and areas of the construction site in a bid to promote sustainable direction on a long-term basis.

Green buildings play an essential role in the reduction of greenhouse gas (GHG) emissions, particularly carbon dioxide by incorporating environmentally favorable design, materials and practices into their design and construction. The development of global trends towards sustainable construction has led to the formation of such institutions as the World Green Building Council to assist national activities in over 100 countries in the harmonisation and development of the sustainable building practice.

Nonetheless, most developing countries are yet to break ground on the issue although the industrialized countries have taken the green building concept into the mainstream building. They are lack of recognition by the stakeholders, proper regulatory structures, and fear of the cost implication of changing over to conventional methods of construction. Other areas are still in the process of instituting measures to achieve the same although barriers are being experienced in developing sound assessment regimes that are locally suited.

The positive impacts of green building do not only consist of an environment-related one, but there are also financial and economic returns, besides the health of the occupants, including decreased operation and maintenance expenses. Different strategic models are incorporated in order to achieve these advantages, such as rating systems, sustainable selection of materials, transportation planning, energy management, and waste reduction [25].

3.7.1 Rating Systems for Sustainable Construction

Sustainability in buildings is now standardized and measured by many green rating systems that have been produced around the world. These models came about due to the rising consumer demand, globalization and urgent need to develop sustainably. Rating systems deliver visible, quantified and verifiability standards in the evaluation of environmental performance of building types.

Such systems as LEED (Leadership in Energy and Environmental Design) and BREEAM (Building Research Establishment Environmental Assessment Method) are the most popular systems that are widely used in various regions. Other nations, like India, have organized national-specific schemes like IGBC to meet the priorities of its country.

Common factors assessed by the rating systems are the efficiency of water, indoor air quality, energy consumption, materials and resources, waste management and amount of pollution. Although the significant benefits of these systems are, among others, the promotion of innovation and distinct performance metrics, they have also been criticized regarding the employment of peculiarities of global location or climate in many cases.

3.7.2 Construction Process Emission

Construction industry is one of the significant contributors to the GHG's worldwide. The points of emissions are various, such as the production of the material needed, the construction stage on the building site, waste management, and transportation planning. Worldwide, the construction sector consumes about 40% of resources and 50% of energy to build structures in certain parts of the developing world. In certain countries specifically the United Kingdom, the industry forms almost 9% of the GHGs of the country.

In this regard, sustainable construction practices should be incorporated all through the design to the implementation. This entails maximizing the efficient use of materials, enhanced construction efficiency, low-carbon technology, and better site logistics all to reduce emissions throughout the project life-cycle.

3.7.3 Sustainable Construction Materials

The materials used in building are critical in the future behavior of buildings in terms of the effect they cause to the environment. Most traditional materials, especially the high embodied energy ones e.g. cement and steel, are energy-consuming to manufacture, thus causing high GHG emissions.

Carbon footprints can be substantially lowered by the construction of environmental friendly materials, recycled content, low-carbon cement, renewable natural materials, bamboo, timber, straw, adobe. What is less known is that these materials, in most cases, consume less energy to manufacture and the bonus is that they sequester carbon as they grow.

New technologies in the production of low-carbon cement involve the addition of magnesia, in order to enable the material to absorb carbon dioxide, during hardening. Efficient use of resources together with green production of concrete can decrease not only the emissions but other resources consumption such as energy and water. Additionally, sustainable materials can really lead to reduced construction waste and extended durability in the life cycle of the structure.

3.7.4 Delivery System/Logistics

Another significant source of carbon emanating almost all the construction activities is the transportation of materials, equipment, and labor. Reduction in transportation emissions can be done by ensuring that materials are sourced locally, efficiency in delivery timing, ensuring just-in-time delivery logistics over idle time and unnecessary movement of vehicles.

Fuel consumption and the carbon footprint of construction works can be dramatically reduced, as well, through the decrease in the number of vehicle trips and the enhancement of the coordination level among the suppliers. Focusing on local purchasing would not only benefit local economies but also help to make it less polluting as regards transport.

3.7.5 Energy Management at Construction Sites

The need to use energy efficiently during the construction leading to minimization of carbon emission is very crucial. There are a number of strategies to be used:

- they are not supposed to employ generator types that are too large unless it would be out of necessity.
- Using energy management plans at the early project planning.
- The prevention of idling of the construction machines and vehicles.
- The implementation of the smart energy systems to control the level of electricity consumption on-site.
- Grid power as opposed to diesel isolated generators.
- Employment of energy efficient lighting, heating as well as the construction equipment.

Such interventions reduce the discharges not only but they also help in terms of cost-saving and keeping the project operational.

3.7.6 The Principle of 3Rs and Waste Management

The other culprit associated with the construction industry that causes environmental degradation and GHGs occurrence is waste. The 3Rs approach will reduce volumes of waste and related emissions drastically by following the Reduce, Reuse, and Recycle.

Material estimation, no over allowing, re-use of offcuts and scrap materials are some of the effective measures to decrease the waste. Creating more sustainability is possible through the usage of recycled materials such as fly ash, slag or reclaimed timber. Diversion of such discarded materials through proper management of waste materials will result in adoption of resource efficiency, preservation of natural ecosystems, and the improvement of green credentials of any construction operation.

Green buildings integrate sustainable strategies that optimize the construction lifecycle (design, materials selection, construction processes, logistics, and operations) ensuring that they help form the backbone of a worldwide strategy to mitigate carbon emissions and fight anthropogenic climate change. The move to low-carbon construction should not only be technologically innovative, but it will make an equally cultural, institutional and policy paradigm shift across the industry.

4 Case Studies

The construction industry in India is slowly coming in line with the world around with its futuristic policies, technology and environmental friendly practices. One of the most important areas of concern has been minimization in carbon emissions during construction of building and infrastructure development. This covers the life cycle of buildings including material choice and energy use throughout the life cycle to the end-of-life plan. Public and private sector actors along with academia are working together more closely now to accept low-carbon approaches and advocate green building certification systems like GRIHA, IGBC and LEED India. Such efforts are leading to the transformation of traditional building towards a more sustainable solution built with an eye to conservation of energy, resources and environmental stability.

One big stride towards this is through the LCA, which has now been advocated as an essential alternative of gauging environmental effects of construction projects. LCA will allow engineers and architects to determine major points of reducing carbon by evaluating the entire life cycle of the production of these items including extracting raw materials, its transportation, demolition and disposal. A number of India pilot projects have been able to triple down on LCA frameworks, returning measurable carbon footprint reductions by using recycled aggregates, alternative binders, and optimised concise constructions.

Moreover, government intervention through subsidies and incentives to use low-carbon technology is significant in making this shift in construction rapid towards sustainable construction. To give an example, the Ministry of New and Renewable Energy (MNRE) is providing subsidies to install solar photovoltaic systems, energy efficient equipment in the form of public and private infrastructure, and waste-to-energy conversion system. Such subsidies lower the first cost threshold and stimulate extensive diffusion of the renewable energy system, particularly in the urban development and smart city initiatives.

Academically, there is an increasing demand of incorporation of the aspect of sustainability into the Indian technical institutions curriculum in civil engineering. Top universities like IITs and NITs have introduced optional and mandatory course on green building technologies, climate responsive architecture, LCA, sustainability urban planning. This will see to it that the future generation of civil engineers not only do they understand the issues of sustainability but also have practical tools and framework to put low-carbon solutions in real projects.

Along with the technical and academic developments and changes, the idea of stakeholder awareness programs is also being actively marketed, so as to ensure that the sustainable practices are implemented in a unified way across the sector. The sensitization of the contractors, builders, architects, and end-users concerning the advantages of green construction take place through workshops, seminars, and end-user campaigns by the bodies such as the Bureau of Energy Efficiency (BEE) and the Indian Green Building Council (IGBC). Such programs can help focus on the

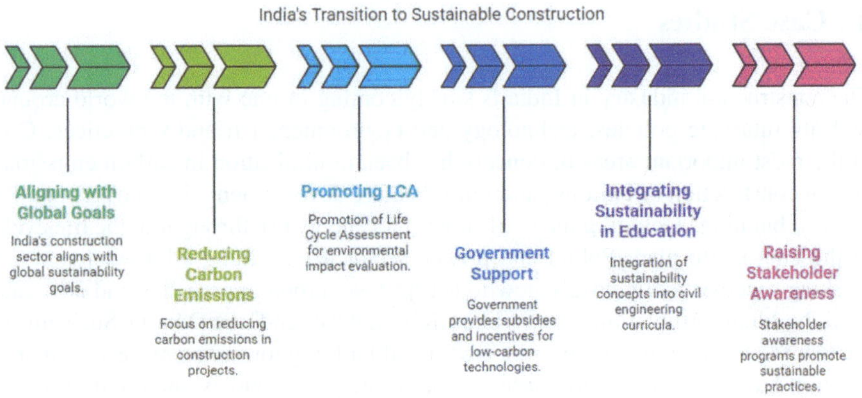

Fig. 13 India's transition to sustainable construction

economic, environmental benefits in the long run by being eco-friendly and setting a culture of sustainability in the ecosystem of construction.

All these together indicate a collective initiative towards making the construction industry in India entirely decarbonized. With its action to make LCA mandatory in large scale projects, provide policy-related support, create a transformation of education, and nourish stakeholder awareness, India is gradually opening a route towards low carbon, resilient, and environmentally sustainable built environment. India's transition to sustainable construction is observed in Fig. 13.

5 Challenges and Barriers

Even though the global society perceived the importance of sustainable development, the construction sector still encounters a number of barriers and challenges that do not support the penetration of green technologies and sustainable conduct among the development methods. The significant initial price of adoption of the sustainable construction solutions takes place as one of the most significant problems. The use of eco-friendly building materials, eco-efficient systems, modern building technologies, and full sustainability evaluations can in fact demand much more upfront investment than the traditional constructing procedures. Such financial obstacles are a significant deterrent to many developers and contractors, particularly in cost-sensitive areas or those with a low-income population, despite obvious long-term operational savings and environmental benefits in operation.

The other significant challenge is the absence of awareness and education on the sustainable practices in construction industry. Most stakeholders, especially the small and medium size enterprises (SME) and local contractors lack much information or are just ignorant of the sustainable economic and environmental benefits of green buildings in the long run. Such lack of knowledge leads to the omission of opportunities to carry out cost effective and environmentally responsible interventions. A poor level of access to case studies, technical advice and training schemes also compounds this problem and makes sustainable construction unable to be mainstream in some markets.

What further compounds the situation are regulatory and policy gaps. Green building standards, energy codes, and environmental regulations are not enforced in a uniform or even a consistent manner in most countries or even in the entire world. In the absence of effective and binding regulations, developers will not feel the need or necessity to develop sustainable ways of undertaking their projects. In some cases where regulations are in place, lack of smooth flow of approval systems, regularized certification environment, and monitoring systems make the process of compliance tedious and uninspiring.

Another obstacle is the lack of available professionals with the knowledge and training of sustainable construction methods. The workforce should have some specialized knowledge and skills since it is required to implement low-carbon materials, renewable energy systems and innovative building technologies. Nevertheless, the existing educational and vocational training curricula in most areas fail to equip engineers, architects, and laborers with sufficient preparation to the green change. Consequently, the industry has difficulty in responding to the increasing market based on sustainable expertise resulting in project delays or project design trade-offs or even rejection of sustainable alternatives because of difficulty in carrying out the project.

Besides that, the structure of construction industry in itself is fragmented posing structural troubles. Constructive projects are undertaken by two or more stakeholders including architects, engineers, contractors, suppliers and regulators whose stakeholder may be working individually with contrasting priorities. Loss of focus in the effort to minimize environmental impact may occur through dilution or complete neglect until good communication, coordination and mutual sustainability goals have been achieved. Such a disconnect kills the efficacy of any even benign green initiative.

Collaborative social engineering that consists of governments, industry leaders, educational institutions, and communities has the approach to overcome these challenges in a multi-dimensional, and holistic perspective. Economic incentives, public-private partnerships, curriculum changes, visibility actions and better policy models will always be key to an upgraded move to a greener construction industry. The act of overcoming these obstacles not only helps in climate targets but also leads to innovation and employment and a better life of ourselves and the generation to come. Challenges in sustainable construction adoption is mentioned in Fig. 14.

6 Policy and Regulatory Frameworks

India has worked out a complex policy and regulation environment to create sustainable construction and minimised the carbon footprint of the built environment. One of the pillars of this initiative is the National Building Code (NBC) of India 2016 which

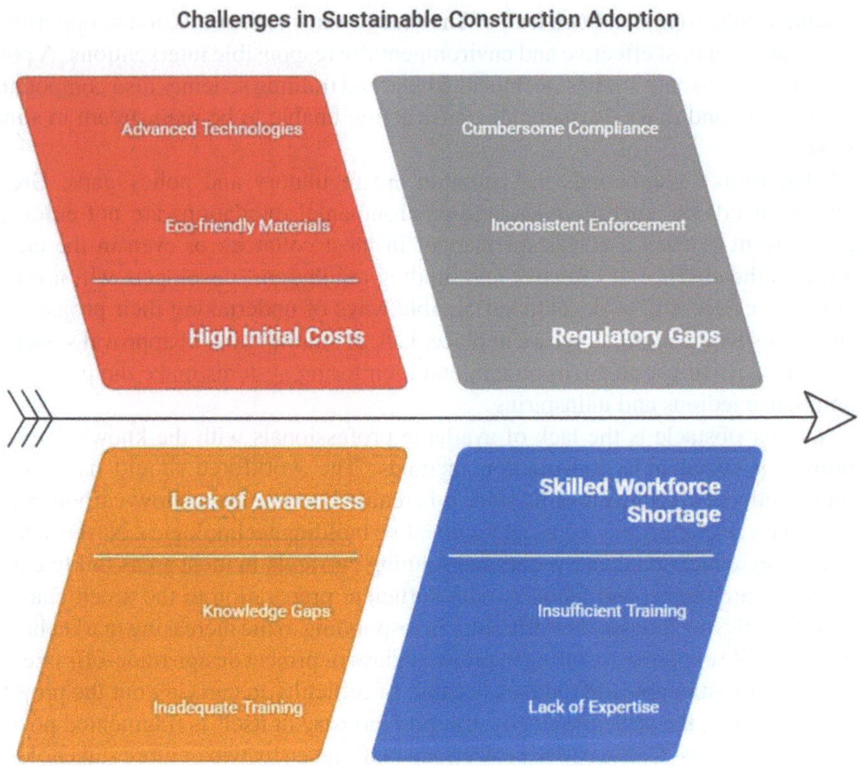

Fig. 14 Challenges in sustainable construction adoption

offers new and modern instructions on how to guarantee safety, sustainability as well as energy efficiency within the building projects. The NBC incorporates climate responsive designing techniques, encourages the use of sustainable resources, and stirs the need to reduce the overall demand on energy consumption in terms of passive and active building design. It acts as a guiding guideline to architects, engineers, constructors and regulation authorities in developing and shaping environmentally sound development buildings.

One of the significant moves towards energy efficiency in commercial buildings is Energy Conservation Building Code (ECBC) which is a code that was developed by an institution in the Ministry of Power called Bureau of Energy Efficiency (BEE). ECBC has prescribed the mandatory energy performance standards on buildings having a connected load of 100 kW or greater or a contract demand of 120 kVA or greater. The code also sets rules on how design can be energy efficient majorly including building envelope, lighting, heating, ventilation and air-conditioning (HVAC) systems, electrical systems and integration with renewable energy. At this level, the ECBC has a significant role to play in determining the climate objectives and energy security of the country, by cutting down on the amount of energy used in operations.

To promote the faster pace of moving towards green construction in India, there has also been the promotion of a green building certification systems especially the Green Rating for Integrated Habitat Assessment (GRIHA) and the Indian Green Building Council (IGBC) ratings. Those structures which are certified through these frameworks are promptly treated in many ways including tax advantages, subsidies and lower stamp duties as well as faster environmental approval. The goal of these incentives is to encourage developers to go an extra mile in embracing energy-efficient technologies and sustainable construction principles implementing not only those that are required.

Besides, the Smart Cities Mission as well as the Sustainable Habitat Mission under the National Action Plan on Climate Change (NAPCC) is another indication of the government that will remain hygienically conscious of the development of the urban centers in the long term. These missions involve support of low-carbon infrastructure, waste management systems, water conservation and incorporation of renewable energy into city planning. They also promote the application of digital technologies and tools monitoring performance in order to provide transparency and constant improvement in terms of sustainability conduct.

The state-level initiatives also complement the efforts taken by such regulatory bodies as many states have come up with ECBC-compliant building bye-laws and have set up special agencies or cells within the urban local authorities to implement enforcement. It also organizes training programs, workshops, and capacity building programs to impart code compliance and green construction skills to the professionals and governmental officials on regular basis.

Conclusively, the emerging policy and regulation environment in India is a very important tool to guide the construction sector to a more permanent and energy-efficient solution. The country has established an excellent framework to make its urban infrastructure environment friendly as it covers mandatory codes with voluntary rating systems and financial incentives. The ongoing enforcement, the regular revision of the codes, and stakeholder investments will be a decisive factor to maintain the relevance and effectiveness of these frameworks on the changing climate and urbanization pressure. India's sustainable construction policy framework is observed in Fig. 15.

Although Indian codes of construction at the national and state level offer a solid basis of sustainable construction, it is becoming more crucial to extend to the global systems and new trends of digital mechanisms of governance. Collaboration of international partners through, among other mechanisms, the United Nations Framework Convention on Climate Change (UNCCC), the Paris Agreement, and regional agreements promote sharing of best practices and access to climate finance mechanisms to finance green infrastructure projects. Such partnerships have helped the nation compare its environmental targets to the standards in the other countries and attracted investment in green urban development.

The second main aspect of sustainable policy formulation is the lifecycle approach to construction regulation. Most regulatory frameworks have always been based

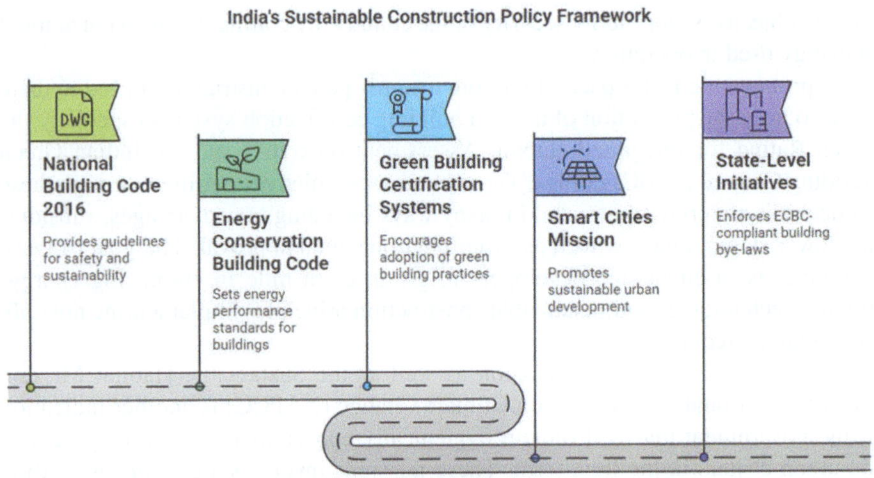

Fig. 15 India's sustainable construction policy framework

on operational energy efficiency but in the present context, there is a tendency to look towards a more comprehensive model that considers embodied carbon during extraction, manufacturing, delivery, and deconstruction/reuse of materials. Too much carbon The LCA tools and standards of carbon accounting have thus become essential tools in the present day building code and environmental impact assessment. The application of these tools into the policy systems can help stakeholders in developing data-based decisions that will lead to the minimization of overall emissions of carbon.

Nevertheless, there are problems regarding application and adherence. The effectiveness of even well-devised policies can be stunted by limited institutional capacity, differing levels of compliance in different jurisdictions and levels of awareness by developers. Further, the enforcement mechanisms are still being developed especially at Tier 2 and Tier 3 cities where the urbanization rate is high, however the structure of governance infrastructure system is weak. These questions can be solved by means of greater inter-agency collaboration and funding of regulatory institutions and the development of performance-based monitoring structures.

The ever-greater emergence of digital technologies promises solutions to these barriers to implementation. Urban planning is moving toward the use of Geographic Information Systems (GIS), Building Information Modeling (BIM), and Internet of Things (IoT) to improve the transparency of our cities and optimize energy consumption alongside real-time feedback on indicators of sustainability. Already, some municipalities have started using digital dashboards to monitor the green building approvals, energy performance, and waste generation hence simplify the process of compliance with the environmental factors within their jurisdiction.

Moreover, the involvement of the population and stakeholders plays a very significant role in the success of regulatory efforts. The act of citizens in relation to sustainable development can be catalyzed by accountability through creation of awareness education, community based monitoring, and open data portals. The gap between policy intent and practical implementation can also be bridged by collaborative governance models contributed by industry associations, academic institutions and civil society organizations.

Finally, the promotion of a policy and regulations in sustainable construction can no longer remain within its previously established parameters. Through incorporation of lifecycle based standards, e-tools, institutional capacities, and multistakeholder involvement, India can consolidate its role in the vanguard of green urbanization and achieve longer-term congruency with both domestic and international climate-related policies.

7 Recommendations

Over the past couple of years, the world battle against climate change has been intensifying its focus on the construction business as well since it significantly contributes to global greenhouse gases—nearly 38% of all globally generated carbon dioxide (CO_2) emissions partially produced by the construction sector. There are two major sources of these emissions which are the embodied carbon and the operation carbon. Embodied carbon encompasses the emission created in the material extraction, production, and building construction procedures whereas operational carbon involves the amount of emission that is produced during the life of the building including heating, cooling, lighting, and building maintenance. The recognition of the overall effect of the two types of carbon is critical in coming up with effective mitigation measures in the built environment.

The construction industry needs an embodied carbon reduction to decrease the construction carbon footprint, and so the embodied carbon should come under consideration first. This can be divided into giving a priority to low-carbon materials, supporting the use of recycled and locally-sourced material, and also promoting construction methods that would reduce waste and energy use. An example is the possibility to reduce emissions by changing cement to fly ash based substitutes or by using engineered wood. Also, designers and engineers are to be prompted to adhere to structural optimization methods and modular building to decrease the consumption of materials further and a carbon intensity.

Simultaneously, it should be ensured that there is an emphasis on operating carbon minimum. This entails developing of structures that use less energy and clean energies. The energy required to heat and cool normal interiors can be slashed using passive design strategies, e.g. correct building orientations, adequate insulation, natural ventilation and daylighting methods. Moreover, the combination of smart energy systems, high-efficiency appliances, and renewable energy such as solar panels and geothermal heat pumps can greatly reduce the emissions of a building during the service life down to irreplaceable values.

The implementation of whole-life carbon assessment tools is needed to be able to make sure that there is a reduction that is comprehensive in all the stages of construction with regard to emissions. These tools help the professionals assess environmental impact at the design stage to when the building is being demolished, and thus more sustainable decisions can be made. By applying such tools to the project at the early stage of the project life cycle, the stakeholders will be able to identify hotspots of high emissions, and they can then choose the strategies which will balance between performances, their affordability and sustainability. Other promising technologies to lower embodied/operational carbon are green roofs, prefabricated building components, and 3D printing.

Finally, cooperation between the public and the private sector is required to increase the speed. Governments should enhance regulation mechanisms through the requirement of carbon disclosures, establishment of performance standards and provision of reward to construction of green buildings. The industry stakeholders (developers, contractors, architects, policy-makers) are encouraged to collaborate to disseminate best practices, procure the green certification and engage in continuous improvement. A concerted effort and change in the thinking to a sustainable position may mean that the construction industry can make a positive difference in the overall carbon reduction targets of the world and add important credence to building a resilient, low-carbon future. Reducing construction's carbon footprint is observed in Fig. 16.

8 Conclusion

The construction industry no longer has an option of reducing its carbon footprint since its representation accounts to a huge portion of the global greenhouse gas emissions. The only way that meaningful carbon reductions are possible is through a holistic, life-cycle assessment, one that focuses on reducing emissions throughout the process, conceptual design and materials, construction, operation and eventual demolition or repurposing. Such a change needs to reconsider established ways of doing things and to accept sustainability as a key underlying principle instead of being an accessory.

Another of extremely significant points is switching to low-carbon sustainable materials like fly ash-based concrete, bamboo, recycled steel, or geopolymer substitutes. These materials not only are embodied carbon reducing materials but also promote the principles of a circuit economy. In conjunction, there is a need to incorporate more intelligent design practices in the construction sector where less material is utilized, waste is lowered, and structures are more efficient. Life cycle assessment (LCA) and Building Information Modeling (BIM) tools are powerful solutions to help make data-driven architecture decisions and make eco-friendly design decisions early in the project lifecycle.

It is also of significance considering the optimization of consumed energy during the working period of buildings. Rainwater harvesting, building orientation, and

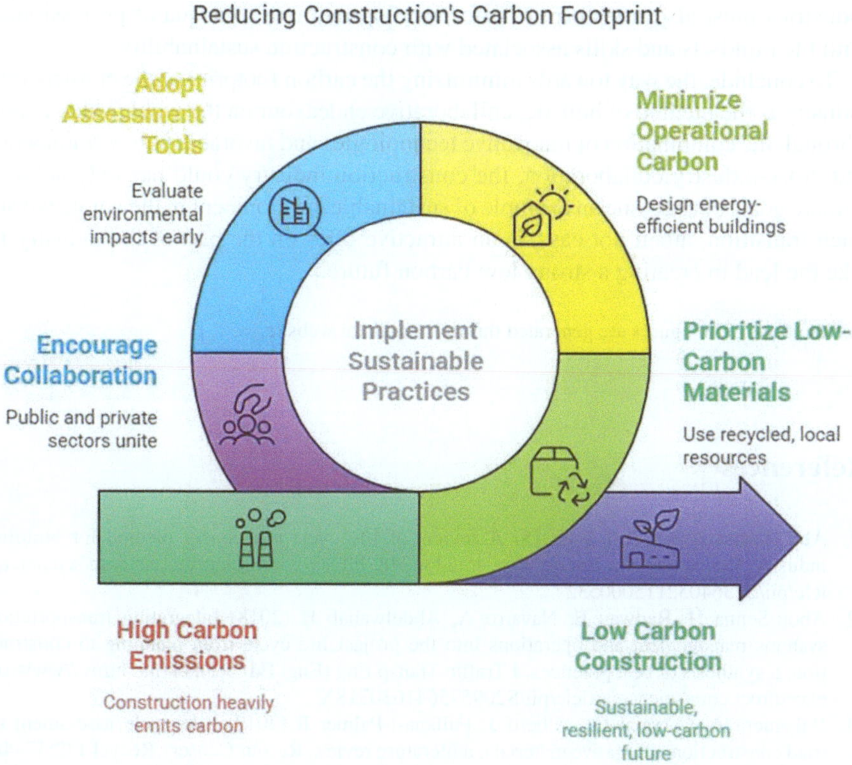

Fig. 16 Reducing construction's carbon footprint

layout, daylighting and passive solar energy design, high-performance insulation, energy-efficient HVAC systems and renewable energy resources including solar panels and geothermal heating are able to significantly reduce operational carbon emission. Finally, the real-time energy consumption and occupant behavior can also be enhanced with the intelligent tools of building management system with energy monitoring, which results in the long-term sustainability.

In order to raise the levels of these strategies, there should be good policy frameworks and government incentives. Carbon taxes, requirements to have environmental certifications, and subsidies on green technologies act as regulatory mechanisms that can support the industry in making it cleaner. The building codes must be updated to require low-carbon construction techniques; urban planning policies must also allow compact, transit oriented, resource efficient construction.

Importantly, transition towards the low-carbon construction industry demands the involvement of all participants in the industry, that is, architects, engineers, contractors, policymakers, material suppliers and clients. The key principles that achieve the elimination of technical, financial, and institutional barriers are cross-disciplinary communication and shared responsibility. Schools and other training facilities in the industries must also transform to offer training required to acquaint professionals with the mindsets and skills associated with construction sustainability.

To conclude, the way towards minimizing the carbon footprint of the construction industry is the method of holistic, collaborative endeavour on the whole value chain. Through the combination of innovative technologies and favorable policy framework and cross-industry collaboration, the construction industry could not only achieve climate goals but become an example of sustainable development in the whole world. Such transition, albeit not easy, is an attractive offer on the part of the industry to take the lead in creating a strong low-carbon future.

Declarations The figures are generated through napkin.ai website.

References

1. Abd Rashid AF, Yusoff S (2015) A review of life cycle assessment method for building industry. Renew Sustain Energy Rev 45:244–248. https://www.sciencedirect.com/science/article/pii/S1364032115000532
2. Abou-Senna H, Radwan E, Navarro A, Abdelwahab H (2018) Integrating transportation systems management and operations into the project life cycle from planning to construction: a synthesis of best practices. J Traffic Transp Eng (Engl Ed) 5(1):44–55. https://www.sciencedirect.com/science/article/pii/S209575641630318X
3. Balaguera A, Carvajal GI, Albertí J, Fullana-i-Palmer P (2018) Life cycle assessment of road construction alternative materials: a literature review. Resour Conserv Recycl 132:37–48. https://www.sciencedirect.com/science/article/pii/S092134491830003X
4. Bayram B, Greiff K (2023) Life cycle assessment on construction and demolition waste recycling: a systematic review analyzing three important quality aspects. Int J Life Cycle Assess 28(8):967–989. https://doi.org/10.1007/s11367-023-02145-1
5. Butera S, Christensen TH, Astrup TF (2015) Life cycle assessment of construction and demolition waste management. Waste Manag 44:196–205. https://www.sciencedirect.com/science/article/pii/S0956053X15300283
6. Cavaliere P (2019) Electrolysis of iron ores: most efficient technologies for greenhouse emissions abatement. In: Cavaliere P (ed) Clean ironmaking and steelmaking processes. Springer International Publishing, pp 555–576. https://doi.org/10.1007/978-3-030-21209-4_10
7. de Brito J, Silva A (2020) Life cycle prediction and maintenance of buildings. In: Buildings, vol 10, no 6. MDPI, p 112. https://www.mdpi.com/2075-5309/10/6/112
8. de la Fuente T, Athanassiadis D, Gonzalez-Garcia S, Nordfjell T (2017) Cradle-to-gate life cycle assessment of forest supply chains: comparison of Canadian and Swedish case studies. J Clean Prod 143:866–881. https://www.sciencedirect.com/science/article/pii/S0959652616320893
9. Enshassi A, Kochendoerfer B, Al Ghoul H (2016) Factors affecting sustainable performance of construction projects during project life cycle phases. Int J Sustain Constr Eng Technol 7(1):50–68. https://penerbit.uthm.edu.my/ojs/index.php/IJSCET/article/view/1195
10. Gava O, Bartolini F, Venturi F, Brunori G, Zinnai A, Pardossi A (2018) A reflection of the use of the life cycle assessment tool for agri-food sustainability. Towar Sustain Glob Food Syst 211. https://library.oapen.org/bitstream/handle/20.500.12657/39485/Towards_Sustainable_Global_Food_Systems.pdf?sequence=1#page=220
11. Gillingham K, Stock JH (2018) The cost of reducing greenhouse gas emissions. J Econ Perspect 32(4):53–72. https://www.aeaweb.org/articles?id=10.1257/jep.32.4.53

12. Guignot S, Touzé S, Von Der Weid F, Ménard Y, Villeneuve J (2015) Recycling construction and demolition wastes as building materials: a life cycle assessment. J Ind Ecol 19(6):1030–1043. https://doi.org/10.1111/jiec.12262
13. Hatfield-Dodds S, Schandl H, Newth D, Obersteiner M, Cai Y, Baynes T, West J, Havlik P (2017) Assessing global resource use and greenhouse emissions to 2050, with ambitious resource efficiency and climate mitigation policies. J Clean Prod 144:403–414. https://www.sciencedirect.com/science/article/pii/S0959652616322338
14. Horobet A, Belascu L, Radulescu M, Balsalobre-Lorente D, Botoroga C-A, Negreanu C-C (2024) Exploring the Nexus between greenhouse emissions, environmental degradation and green energy in Europe: a critique of the environmental Kuznets curve. Energies (19961073) 17(20). https://www.researchgate.net/profile/Alexandra-Horobet/publication/384728379_Exploring_the_nexus_between_Greenhouse_Emissions_Environmental_Degradation_and_Green_Energy_in_Europe_-_A_Critique_of_the_Environmental_Kuznets_Curve/links/67701adf894c552085315a9b/Exploring-the-Nexus-between-Greenhouse-Emissions-Environmental-Degradation-and-Green-Energy-in-Europe-A-Critique-of-the-Environmental-Kuznets-Curve.pdf
15. Hryhorovskyi P, Chukanova N, Murasova O, Chernenko K (2023) Analysis of the duration of operational stage as a part of constructed assets life cycle with impact of natural and man-made factors. AIP Conf Proc 2490(1):040002. https://pubs.aip.org/aip/acp/article-abstract/2490/1/040002/2928219
16. Huang B, Gao X, Xu X, Song J, Geng Y, Sarkis J, Fishman T, Kua H, Nakatani J (2020) A life cycle thinking framework to mitigate the environmental impact of building materials. One Earth 3(5):564–573. https://www.cell.com/one-earth/fulltext/S2590-3322(20)30540-6
17. Hyder Z, Ripepi NS, Karmis ME (2016) A life cycle comparison of greenhouse emissions for power generation from coal mining and underground coal gasification. Mitig Adapt Strat Glob Chang 21(4):515–546. https://doi.org/10.1007/s11027-014-9561-8
18. Jacob C, Nandra A, Gupta J (2025) Strategic concepts, challenges, and life-cycle assessment for sustainable construction and building circularity in the real estate sector. Circ Econ Sustain 5(2):1201–1217. https://doi.org/10.1007/s43615-024-00466-3
19. Jonas M, Bun R, Nahorski Z, Marland G, Gusti M, Danylo O (2019) Quantifying greenhouse gas emissions. Mitig Adapt Strat Glob Chang 24(6):839–852. https://doi.org/10.1007/s11027-019-09867-4
20. Kamali M, Hewage K (2016) Life cycle performance of modular buildings: a critical review. Renew Sustain Energy Rev 62:1171–1183. https://www.sciencedirect.com/science/article/pii/S1364032116301411
21. Kovacic I, Zoller V (2015) Building life cycle optimization tools for early design phases. Energy 92:409–419. https://www.sciencedirect.com/science/article/pii/S0360544215003217
22. Labaran YH, Mathur VS, Muhammad SU, Musa AA (2022) Carbon footprint management: a review of construction industry. Clean Eng Technol 9:100531. https://www.sciencedirect.com/science/article/pii/S2666790822001367
23. Lamb WF, Wiedmann T, Pongratz J, Andrew R, Crippa M, Olivier JG, Wiedenhofer D, Mattioli G, Al Khourdajie A, House J (2021) A review of trends and drivers of greenhouse gas emissions by sector from 1990 to 2018. Environ Res Lett 16(7):073005. https://iopscience.iop.org/article/10.1088/1748-9326/abee4e/meta
24. Li H, Deng Q, Zhang J, Olubunmi Olanipekun A, Lyu S (2019) Environmental impact assessment of transportation infrastructure in the life cycle: case study of a fast track transportation project in China. Energies 12(6):1015. https://www.mdpi.com/1996-1073/12/6/1015
25. Li X, Wang C, Kassem MA, Wu S-Y, Wei T-B (2022) Case study on carbon footprint life-cycle assessment for construction delivery stage in China. Sustainability 14(9):5180. https://www.mdpi.com/2071-1050/14/9/5180
26. Li X-J, Zheng Y (2020) Using LCA to research carbon footprint for precast concrete piles during the building construction stage: a China study. J Clean Prod 245:118754. https://www.sciencedirect.com/science/article/pii/S0959652619336248

27. Lu K, Jiang X, Tam VW, Li M, Wang H, Xia B, Chen Q (2019) Development of a carbon emissions analysis framework using building information modeling and life cycle assessment for the construction of hospital projects. Sustainability 11(22):6274. https://www.mdpi.com/2071-1050/11/22/6274
28. Lu Y, Song X (2017) Beyond boundaries: a global use of life cycle inventories for construction materials. J Clean Prod 156:876–887. https://www.sciencedirect.com/science/article/pii/S0959652617307060
29. Luo L, Chen Y (2020) Carbon emission energy management analysis of LCA-based fabricated building construction. Sustain Comput Inform Syst 27:100405. https://www.sciencedirect.com/science/article/pii/S2210537920301323
30. Marini A, Passoni C, Belleri A (2018) Life cycle perspective in RC building integrated renovation. Procedia Struct Integr 11:28–35. https://www.sciencedirect.com/science/article/pii/S2452321618301069
31. Nadoushani ZSM, Akbarnezhad A (2015) Effects of structural system on the life cycle carbon footprint of buildings. Energy Build 102:337–346. https://www.sciencedirect.com/science/article/pii/S0378778815300190
32. Nägeli C, Farahani A, Österbring M, Dalenbäck J-O, Wallbaum H (2019) A service-life cycle approach to maintenance and energy retrofit planning for building portfolios. Build Environ 160:106212. https://www.sciencedirect.com/science/article/pii/S0360132319304226
33. Pal SK, Takano A, Alanne K, Siren K (2017) A life cycle approach to optimizing carbon footprint and costs of a residential building. Build Environ 123:146–162. https://www.sciencedirect.com/science/article/pii/S036013231730286X
34. Pizzol M, Laurent A, Sala S, Weidema B, Verones F, Koffler C (2017) Normalisation and weighting in life cycle assessment: quo vadis? Int J Life Cycle Assess 22(6):853–866. https://doi.org/10.1007/s11367-016-1199-1
35. Ram VG, Kishore KC, Kalidindi SN (2020) Environmental benefits of construction and demolition debris recycling: evidence from an Indian case study using life cycle assessment. J Clean Prod 255:120258. https://www.sciencedirect.com/science/article/pii/S095965262030305X
36. Ritchie H, Roser M (2017) CO_2 and other greenhouse gas emissions. Our World in Data. https://www.sesync.org/sites/default/files/case-studies/Evidence%20of%20Global%20Warming%20in%20VA%20Case%20Study%20%E2%80%93%20Supplemental%20Document%202.pdf
37. Sanchez-Londono D, Barbieri G, Fumagalli L (2023) Smart retrofitting in maintenance: a systematic literature review. J Intell Manuf 34(1):1–19. https://doi.org/10.1007/s10845-022-02002-2
38. Saxe S, Kasraian D (2020) Rethinking environmental LCA life stages for transport infrastructure to facilitate holistic assessment. J Ind Ecol 24(5):1031–1046. https://doi.org/10.1111/jiec.13010
39. Schwartz Y, Raslan R, Mumovic D (2018) The life cycle carbon footprint of refurbished and new buildings—a systematic review of case studies. Renew Sustain Energy Rev 81:231–241. https://www.sciencedirect.com/science/article/pii/S1364032117311309
40. Shurpali N, Agarwal AK, Srivastava V (eds) (2019) Greenhouse gas emissions: challenges, technologies and solutions. Springer Singapore. https://doi.org/10.1007/978-981-13-3272-2
41. Singh S, Babbitt C, Gaustad G, Eckelman MJ, Gregory J, Ryen E, Mathur N, Stevens MC, Parvatker A, Buch R, Marseille A, Seager T (2021) Thematic exploration of sectoral and cross-cutting challenges to circular economy implementation. Clean Technol Environ Policy 23(3):915–936. https://doi.org/10.1007/s10098-020-02016-5
42. Sizirici B, Fseha Y, Cho C-S, Yildiz I, Byon Y-J (2021) A review of carbon footprint reduction in construction industry, from design to operation. Materials 14(20):6094. https://www.mdpi.com/1996-1944/14/20/6094
43. Van Ruijven BJ, Daenzer K, Fisher-Vanden K, Kober T, Paltsev S, Beach RH, Calderon SL, Calvin K, Labriet M, Kitous A (2016) Baseline projections for Latin America: base-year assumptions, key drivers and greenhouse emissions. Energy Econ 56:499–512. https://www.sciencedirect.com/science/article/pii/S014098831500033X

44. Vieira DR, Calmon JL, Coelho FZ (2016) Life cycle assessment (LCA) applied to the manufacturing of common and ecological concrete: a review. Constr Build Mater 124:656–666. https://www.sciencedirect.com/science/article/pii/S0950061816312259
45. Wuni IY, Shen GQ (2020) Critical success factors for management of the early stages of prefabricated prefinished volumetric construction project life cycle. Eng Constr Arch Manag 27(9):2315–2333. https://www.emerald.com/insight/content/doi/10.1108/ECAM-10-2019-0534/full/html
46. Yang X, Hu M, Wu J, Zhao B (2018) Building-information-modeling enabled life cycle assessment, a case study on carbon footprint accounting for a residential building in China. J Clean Prod 183:729–743. https://www.sciencedirect.com/science/article/pii/S0959652618303767
47. Yuan C, McClure T, Cai H, Dunston PS (2017) Life-cycle approach to collecting, managing, and sharing transportation infrastructure asset data. J Constr Eng Manag 143(6):04017001. https://doi.org/10.1061/(ASCE)CO.1943-7862.0001288
48. Yuan L, Yang B, Lu W, Peng Z (2024) Carbon footprint accounting across the construction waste lifecycle: a critical review of research. Environ Impact Assess Rev 107:107551. https://www.sciencedirect.com/science/article/pii/S0195925524001380
49. Zhang Z, Wang B (2016) Research on the life-cycle CO_2 emission of China's construction sector. Energy Build 112:244–255. https://www.sciencedirect.com/science/article/pii/S0378778815304631

Process Safety Aspects in Low Carbon Hydrogen Production via Electrolysis

Eugênio Paceli Carvalho da Silva, Décio Yukishigue Fujimoto Júnior, and Victor Hugo Souza de Abreu

Abstract Low carbon hydrogen, produced by water electrolysis using renewable energy, stands out as a crucial energy vector for the decarbonization of industrial and transportation sectors. This chapter analyzes the safety risks associated with green hydrogen production, focusing on alkaline and proton exchange membrane (PEM) electrolysis systems, proposing mitigation measures based on technical standards, industrial best practices, and risk management tools. Through a mixed methodological approach—integrating an integrative literature review, comparative analysis of standards (ABNT NBR IEC 60079-10-1 and Petrobras N-2918), application of HAZOP and Bow Tie, and case studies in pilot units, critical challenges were identified, such as the high flammability of hydrogen, explosion risks from mixing with oxygen, material embrittlement (Hydrogen embrittlement—HE), and leaks in pressurized systems. The results revealed that the Petrobras N-2918 Standard, more conservative than the ABNT, offers practical guidelines for hazardous area classification, reducing risks in semi-confined environments. Additionally, the selection of corrosion-resistant materials, the implementation of early detection sensors, and the adoption of periodic inspection protocols (according to ASME B31.12-2023) are essential to ensure operational safety. It is concluded that the viability of green hydrogen depends on the integration of advanced technologies, robust regulations, and proactive risk management, ensuring its contribution to a safe and sustainable energy transition.

Keywords Low carbon hydrogen · Electrolysis · Process safety · Risk assessment

---

E. P. C. da Silva · D. Y. F. Júnior · V. H. S. de Abreu (✉)
MBA in Energies from the Alberto Luiz Coimbra Institute for Graduate Studies and Engineering Research at the Federal University of Rio de Janeiro, Rio de Janeiro, Brazil
e-mail: vhsa@poli.ufrj.br

E. P. C. da Silva
e-mail: eugenio.silva@mbcursos.coppe.ufrj.br

D. Y. F. Júnior
e-mail: decio.fujimoto@mbcursos.coppe.ufrj.br

S. S. Muthu (ed.), *Reduction of Industrial Carbon Footprint*, Environmental Footprints and Eco-design of Products and Processes,
https://doi.org/10.1007/978-3-032-10913-2_5

1 Introduction

Geopolitical concerns associated with the depletion of fossil fuels, environmental pollution issues, and climate change, as well as the need to ensure energy supply independence, make the pursuit of a low carbon economy an essential factor [1, 2]. In this context, hydrogen as an energy source is likely to be inevitable in the coming decades [3, 4].

Hydrogen is a relevant energy vector for the transition towards a low carbon economy, produced from water electrolysis using electricity generated from renewable sources such as wind and solar energy [5, 6]. This clean and versatile fuel has enormous potential to decarbonize industrial, transportation, and power generation sectors [7]. However, the production, storage, and transportation of hydrogen involve significant safety-related challenges [8, 9]. Low carbon hydrogen production can be enabled by different technological routes, including water electrolysis using electricity from renewable sources like solar, wind, and other sustainable matrices. Additionally, steam reforming of methane or biomethane can be employed, provided it is associated with carbon capture and storage systems.

With an energy density of 33 kWh/kg, hydrogen is the most promising element in the energy field. For the same weight, the combustion of hydrogen generates three times more energy than gasoline. Besides having significant energy efficiency, hydrogen combustion does not emit carbon dioxide (CO_2), making its use environmentally neutral [10]. However, despite the environmental advantages, hydrogen production involves highly critical industrial risks that require effective safety actions to protect people, facilities, and the environment. Among the most pressing challenges are hydrogen's high flammability, its wide explosibility range, and the simultaneous presence of oxygen at the same production site.

Moreover, according to Calabrese et al. [11], hydrogen has a transparent flame and being very light, its diffusive nature allows small leaks to spread rapidly in the air over large distances. However, this characteristic is quite positive in terms of risk effect, as hydrogen dispersion in the air is very effective, and it can quickly reduce its concentration. Nevertheless, this represents an explosion risk, as hydrogen concentrations can reach the flammable range very close to the point of origin at the initial moment of gas release.

This chapter aims to identify the main safety risks associated with green hydrogen production through electrolysis, presenting mitigation measures based on best industrial practices and updated research, with the objective of supporting future risk studies to be conducted in low carbon hydrogen plants.

2 Methodology

The study adopted a mixed methodological approach, integrating qualitative and quantitative strategies, structured into four main stages:

- **Integrative literature review**, involving the collection of scientific articles, technical standards (ABNT, ISO, IEC), and sectoral documents (e.g., reports from the European Industrial Gases Association), aiming to contextualize the risks associated with green hydrogen production and electrolysis technologies.
- **Comparative technical analysis** of safety standards, focusing on the classification of hazardous areas (ABNT NBR IEC 60079-10-1 vs. Petrobras N-2918) and criteria for electrolyzer operation (ISO 22734).
- **Application of risk management tools** (Hazard and Operability Study—HAZOP and Bow Tie Analysis) for modeling critical scenarios in hydrogen production and storage systems, including matrices for assessing probability and severity.
- **Practical case studies**, involving the evaluation of a green hydrogen pilot unit, including pipeline inspections (based on ASME B31.12-2023) and area classification according to ABNT and Petrobras standards.

Data collection combined primary sources (e.g., comparative tables for area classification) and secondary sources (e.g., physicochemical data of hydrogen). The results were validated through methodological triangulation, crossing theoretical, normative, and practical evidence in real contexts.

3 Hydrogen Production via Electrolysis: Overview

Water electrolysis is the only process with no carbon dioxide emissions into the atmosphere and produces zero emissions when powered by electricity obtained from renewable sources [12]. Water electrolysis involves passing an electric current through an aqueous solution, typically containing an electrolyte to enhance electrical conductivity, resulting in the release of H (at the cathode) and O (at the anode). The most common technologies include alkaline electrolysis and proton exchange membrane (PEM) electrolysis, operating under different temperature and pressure conditions. Other types of electrolyzers exist but are of lesser commercial relevance.

3.1 Alkaline Electrolysis

The alkaline electrolyzer operates based on electrolytes of potassium hydroxide (KOH) or sodium hydroxide (NaOH) with mass concentrations of 25–30%. Nickel (Ni) and cobalt oxides are used to form the materials for the anode and cathode, respectively.

Additionally, the potassium hydroxide (KOH) electrolyte is the most commonly used, with a concentration of 30–40%, and alkalinity is generated by circulating the electrolyte through both electrodes. Operating temperatures can range from 65 to 100 °C, functioning at pressures from 1 to 30 bar, typically used around 25–30 bar, although there are models that operate at atmospheric pressure and others that can reach up to 448 bar. Other possible electrolytes, such as NaOH or NaCl solutions, are less commonly used. The main and obvious disadvantage of the alkaline electrolyte is its corrosive nature [13].

3.2 PEM Electrolysis

A PEM electrolyzer is a static electrochemical device that converts electrical energy into chemical energy primarily through the dissociation of water (input) into oxygen and hydrogen (output) with the help of direct current (DC) [14].

In PEM electrolyzers, the electrolyte is a thin gas-tight polymer membrane (less than 0.2 mm thick) with a reticulated structure and a strongly acidic character due to the presence of sulfonic acid functional groups (HSO_3^-). These groups are responsible for the proton H^+ conductivity of the materials through an ionic exchange mechanism. As for PEM fuel cells, the most commonly used membrane for water electrolysis is Nafion (a registered trademark of DuPont). The electrodes typically consist of noble metals such as platinum or iridium [13].

Commercial state-of-the-art PEM electrolyzers operate at hydrogen output pressures of 30–40 bar, but there are many end-use applications that require considerably higher-pressure levels, such as injection into the natural gas grid (approximately 70 bar), green methanol production (approximately 80 bar), and ammonia production (approximately 200 bar) [15].

In both cases, the production of large quantities of hydrogen and oxygen in relatively confined spaces reinforces the need for stringent safety systems [5]. Table 1 presents the main characteristics of the electrolyzers.

4 Process Safety in Low Carbon Hydrogen Production

Due to its highly flammable nature and wide flammability range, hydrogen requires stringent measures to prevent risks such as leaks, explosions, and operational failures. Additionally, the presence of high-pressure systems and the need to ensure the purity of the produced gas make robust engineering practices, compliance with international standards, and the use of advanced technologies indispensable.

One of the main objectives of process safety is to prevent the loss of primary containment of hydrocarbons or other hazardous products, avoiding an unplanned release (leak) of these substances. In this sense, safety with hydrogen heavily relies on the design phase of the project and how it is maintained during operation.

A hydrogen leak is very difficult to detect, as it is a colorless, odorless, and tasteless gas; it is also more prone to transitioning from deflagration to detonation than most other flammable gases [4].

Table 1 Comparison of electrolyzers

Parameter	Alkaline	PEM
Electrolyte	Potassium Hydroxide (KOH) Solution—(20–40%)	Perfluorosulfonic acid membrane (PFSA) (e.g., Nafion)
Cathode	Nickel—Ni–Mo Alloys	Platinum, platinum-palladium alloy
Anode	Nickel—Ni–Co Alloys	Ruthenium oxide, iridium oxide
Operating temperature (°C)	60–80	50–80
Operating pressure (bar)	30	70
Current density (A/cm^2)	0.4–1.2	0.6–2.0
Power density (mW/cm^2)	< 1	< 4.4
Cell lifespan	60,000–100,000	20,000–60,000
Technology	Mature	Commercialization
Gas purity (%)	99.50–99.95	99.99
Response time	Seconds	Milliseconds
Advantages	Accessible for large plant projects, low cost, and high lifespan	No corrosive substances, high power density, high pressures
Disadvantages	Low power density, maintenance costs (highly corrosive system)	High cost and rapid degradation
H_2 production (m^3/h)	< 760	Up to 450
Cell voltage	1.8–2.4	1.8–2.2
Cost	USD 500–1400/kW	USD 1100–1800/kW

Source IRENA [16], Green Hydrogen Cost Reduction: Scaling up Electrolysers to Meet the 1.5 °C Climate Goal

4.1 Process Safety

Risk is, by definition, the combination of the probability of an event and its consequence. Danger, on the other hand, is a condition that has the potential to cause accidents. According to MTU [17], safety primarily involves understanding hazards and risks, managing those risks by providing appropriate layers of protection to reduce the frequency and severity of accidents, and learning from accidents when they occur. It involves identifying and understanding potential hazards, assessing consequences, and adding layers of protection (safeguards) to prevent and/or mitigate accidents, thereby protecting people, the environment, and property.

Thus, one of the main objectives of process safety is to prevent the loss of primary containment of hydrocarbons or other hazardous products, meaning to avoid an unplanned release (leak) of these substances. In this sense, safety with hydrogen heavily relies on the design phase of the project and how it is maintained during operation.

4.2 Hydrogen and Oxygen Mixture

The mixture of hydrogen (H_2) and oxygen (O_2) in electrolyzers presents significant risks, primarily related to safety. These gases, when combined in certain proportions, can form a highly flammable and explosive mixture. An explosion within the system can cause catastrophic damage to the components of the electrolyzer, including membrane ruptures, valves, and containment systems. It can also compromise the integrity of nearby facilities and structures, as well as endanger operators and personnel in proximity.

Hydrogen is a fuel with a wide flammability range both in air, from 4 to 75% by volume, and in oxygen, from 4 to 95%. It is also the fuel with the highest energy content per unit mass, with its higher heating value (HHV) being 3.54 kWh/Nm^3 (39.42 kWh/kg) (Rajeswhwar et al. 2008, [13]).

For explosion purposes, in the case of H_2 mixed with O_2, exceeding a few percentage points of the Lower Explosive Limit (LEL) produces a fairly restricted effect because the explosion occurs under conditions of low concentrated fuel (H_2). On the other hand, for O_2 mixed with H_2, the effect is more disastrous because, in this case, H_2, the fuel, is dominant. Therefore, it is recommended to install a detector to interrupt the electrolysis process when the concentration of O_2 mixed with H_2 reaches 1.1% and 1.6%, with 1.1% triggering an alarm and 1.6% indicating process shutdown. Another recommendation is to use pure nitrogen (N_2) before each restart of the electrolyzer to make the plant less reactive by removing H_2 and air from the system, as this will dilute hydrogen-rich mixtures to avoid the risk of explosion [18].

The technical standard ABNT ISO 22734 states that the electrolyzer must have an emergency shutdown function to immediately remove power from the system in the event of a real or imminent danger that cannot be corrected by process control. This emergency shutdown should be triggered when the maximum volume fraction of 2% H_2 in air, 2% H_2 in O_2, or 3% excess O_2 in H_2 is reached. The response time and accuracy of the gas mixture detection control for the emergency shutdown must be considered in the risk analysis, and this should be determined by the manufacturer, who must indicate whether there are hazards of H_2 in air, H_2 in O_2, or O_2 in the fuel gas mixture [19].

4.3 Area Classification

Hydrogen is a very small molecule with a molecular weight of 1.08, being a highly volatile and explosive gas, much lighter than air, which makes it very prone to leaks. According to Herath et al. 2009, cited in Kleszcz and Assadi [20], hydrogen has a low ignition energy threshold, with an explosion volume fraction between 18 and 59%, and its minimum required ignition energy is very low, at only 0.02 mJ, which represents a certain level of risk when there is a leak because it has a substantial potential for explosions and carries a significant amount of explosive energy.

A hydrogen production module integrates various units such as electrolyzers, power supply, tanks, pumps, separators, dryers, compression systems, and storage. Many of these components operate at high pressures and have connections that are susceptible to leaks, increasing the risks. Furthermore, several pieces of equipment may not be certified for use in hazardous areas, making these installations challenging in terms of explosion safety.

To mitigate these risks, it is essential to define, monitor, and control the explosion hazard, as well as to adopt appropriate protection methods for the equipment used. The best way to mitigate the risk of hydrogen explosion is to ensure adequate ventilation. In open environments, where ventilation is naturally efficient, hydrogen dissipates quickly below its lower flammability limit due to its low molecular weight. However, in confined or semi-confined spaces, hydrogen tends to accumulate in the upper areas, creating a risk of forming an explosive mixture. To prevent this, it is essential that the ventilation design of hydrogen modules is capable of providing sufficient air circulation to disperse the gas and prevent the flammability limit from being reached. The classification design of the hydrogen module should always aim for a Zone 2 classification, according to the IEC 60079-10-1* standard, which defines performance criteria for ventilation systems in explosive atmospheres [21].

*In Brazil, its current version is ABNT NBR IEC 60079-10-1 [22], which is in its third edition as of 10/19/2022. Table 2 (Table D.1 of ABNT 60079-10-1—Zones for release degree and ventilation effectiveness) presents the area classifications based on the degree of release and availability of ventilation.

Table 2 defines the ventilation design criteria for a Zone 2 application based on the degree of release. For evaluation, a secondary release is assumed because a primary release of hydrogen in a closed area is never a good idea. To achieve a Zone 2 classification, the ventilation design must provide a criterion for average dilution with some degree of availability [21].

4.3.1 ABNT Standard 60079-10-1 and Petrobras Standard 2918 (N-2918)

A comparative study was conducted by the SENAI Institute of Innovation in Renewable Energies between the standards **ABNT NBR IEC 60079-10-1 and N-2918** with the objective of analyzing divergences in the classification of classified areas for a **pilot project of water electrolysis plant**, employing 2.0 MW PEM technology, where the source of electrical energy will be solar. The plant will be implemented in a Petrobras thermoelectric power plant located in the Northeast region of Brazil.

The N-2918—Explosive Atmospheres—Area Classification—is a public access Petrobras standard that determines the minimum requirements for the preparation of area classification studies in its terrestrial and maritime installations [23].

According to IEC 60079-10, hydrogen belongs to the gas group IIC, which includes the most dangerous gases, characterized by high flammability and low minimum ignition energy. Regarding temperature class, hydrogen has a very high

Table 2 Zones for release degree and ventilation effectiveness

Degree of release	Effectiveness of ventilation						
	High dilution			Medium dilution			Low dilution
	Availability of ventilation						
	Good	Satisfactory	Poor	Good	Satisfactory	Poor	Good, satisfactory or poor
Continuous	Unclassified (Zone 0 ED)[a]	Zone 2 (Zone 0 ED)[a]	Zone 1 (Zone 0 ED)[a]	Zone 0	Zone 0 + Zone 2[c]	Zone 0 + Zone 1	Zone 0
Primary	Unclassified (Zone 1 ED)[a]	Zona 2 (Zona 1 ED)[a]	Zone 2 (Zone 1 ED)[a]	Zone 1	Zone 1 + Zone 2	Zone 1 + Zona 2	Zone 1 + Zona 0[d]
Secondary[b]	Unclassified (Zone 1 ED)[a]	Unclassified (Zone 2 ED)[a]	Zone 2	Zone 2	Zone 2	Zone 2	Zone 1 And even Zone 0[d]

[a] Zone 0 ED, Zone 1 ED, or Zone 2 ED indicate a theoretical zone that may have a negligible extent (ED) under normal conditions

[b] Zone 2 generated by a secondary release may exercise that attributed by primary or continuous releases. In this case, it is advisable to disregard the greater distance

[c] Zone 1 is not necessary in this case, i.e., when the release is not controlled by ventilation, it is necessary to consider the presence of a small Zone 0 surrounded by a larger Zone 2, due to ventilation failure situations

[d] Zone 0 is considered if ventilation is so poor and the release is of such magnitude that, in practice, an explosive gas atmosphere exists continuously, that is, when approaching a condition of non-existence of ventilation

'+'means' surrounded by

The availability of ventilation in enclosed spaces with natural ventilation is not considered 'good'

ignition temperature, around 585 °C, which places it in temperature class T1 (maximum surface temperatures up to 450 °C).

The pilot hydrogen generation unit will consist of:

- Water Electrolyzer of the PEM type (Raw Material: Deionized Water);
- Proton Exchange Membrane Fuel Cell (PEMFC);
- 4 Cylindrical tanks for hydrogen storage.

Below is a summary of how each of the items above in the unit was classified.

Based on the data presented in Table 3, N-2918 is much more restrictive than IEC 60079-1, as N-2918 provides more specific criteria for hydrogen installations, delimiting an area extension of 4.5 m × 7.5 m for both zone 1 and zone 2. Both consider hydrogen as a highly dangerous gas, but N-2918 tends to be more detailed and practical for use in the oil and gas industries. For greater safety of installations,

Table 3 Comparison between IEC 60079-410-1 and N-2918

<table>
<tr><th colspan="2">Electrolyzer</th></tr>
<tr><td colspan="2">The electrolyzer has 3 vent outlets, one from the Hydrogen analyzer, one oxygen outlet from the anode, and one outlet from the oxygen analyzer. For area classification purposes, only the hydrogen outlet was considered. In terms of valves, at the cathode outlet of the electrolysis module, there is 1 PSV valve, 1 PSV valve in the Gas–Liquid separator, and 2 PSV valves in the PSA (Pressure Swing Adsorption) system for H_2 dehumidification, which remain normally closed. All are connected to the same vent</td></tr>
<tr><td>IEC 60079-10-1</td><td>N-2918</td></tr>
<tr><td>The assemblies generate a primary release source, with temperature class T1 and Group IIC, generating a classified area of negligible extension</td><td>He assemblies generate a primary release source, with temperature class T1 and Group IIC, generating a classified area of zone 1 around the electrolyzer container, with extensions of 4.5 m horizontally and 7.5 m vertically</td></tr>
<tr><th colspan="2">Fuel cell</th></tr>
<tr><td colspan="2">Similarly to the electrolyzer, the fuel cell has safety valves and vents, generating a primary release source, with temperature class T1 and Group IIC</td></tr>
<tr><td>IEC 60079-10-1</td><td>N-2918</td></tr>
<tr><td>Área classificada de zona 1 em torno do container da célula combustível de extensão desprezível</td><td>Área classificada de zona 1 em torno do container da célula combustível, com extensões de 4.5 m horizontal e 7.5 m vertical</td></tr>
<tr><th colspan="2">Storage</th></tr>
<tr><td colspan="2">The storage system consists of 4 hydrogen tanks containing 30m^3 each, totaling 120m^3. The storage system has valves through which hydrogen could leak, but this does not occur under normal operating conditions, which characterizes a secondary release source, with temperature class T1 and Group IIC</td></tr>
<tr><td>IEC 60079-10-1</td><td>N-2918</td></tr>
<tr><td>Area classification of zone 2 with negligible extension</td><td>Area classification of zone 2, with extensions of 4.5 m horizontally and 7.5 m vertically</td></tr>
</table>

even if this may imply higher costs, the use of N-2918 for hydrogen installations is recommended.

4.4 *Corrosion and Material Degradation*—Hydrogen Embrittlement *(HE)*

Electrolysis cells, electrodes, and piping are subject to aggressive chemical and electrochemical conditions. Corrosion reduces process efficiency and can lead to structural failures, creating hydrogen leak points [5].

One of the main forms of degradation of materials exposed to hydrogen is embrittlement. While corrosion involves the degradation of material due to chemical or electrochemical reactions with the environment, hydrogen embrittlement occurs when hydrogen atoms penetrate the crystalline structure of the metal, reducing its ductility and mechanical strength without necessarily causing material loss. Hydrogen embrittlement occurs when metals become brittle and develop cracks due to prolonged exposure to hydrogen, leading to degradation of mechanical properties and potential leaks, which can result in environmental and safety risks, especially due to the ignition of hydrogen after rupture. Although the exact mechanisms are not yet fully understood, factors such as hydrogen concentration and purity, pressure, temperature, exposure time, stress state, microstructure, and surface conditions influence the process. Therefore, any repairs or modifications to systems operating with hydrogen must be carefully planned and tested to avoid failures [24].

According to [25], hydrogen diffuses and accumulates in regions under stress, leading to fracture failure when it reaches a critical concentration. Hydrogen sources are classified as internal (generated during processes such as casting, welding, and electroplating) and external (resulting from corrosion, hydrogen gas, or environments with H_2S). Prevention of hydrogen embrittlement (HE) can be achieved through two main approaches: surface treatments and modification of the material's microstructure.

(a) **Surface Treatments**:

- **Surface Coating**: Application of films to suppress hydrogen ingress, increasing resistance to HE.
- **Surface Modification**: Techniques such as nitriding, carburizing, and shot peening improve resistance to HE by stabilizing phases such as austenite and preventing detrimental transitions.

(b) **Microstructure Modification**:

- **Alloying Elements**: Adjusting the chemical composition of the material can reduce HE. For example, decreasing elements such as C, Si, P, and S, or increasing Ni, Al, and Mo improves resistance. Elements such as Mo, V, and Ti form carbides that act as hydrogen traps.
- **Microstructure**: Susceptibility to HE varies according to microstructure: martensite is the most susceptible, followed by bainite, pearlite, and austenite. Increasing the austenite content, which has low diffusion and high hydrogen solubility, reduces HE. Stabilizing austenite, such as with the addition of carbon, also increases resistance.

Hydrogen embrittlement is a significant challenge in materials engineering, affecting the safety and reliability of metallic components. Mitigating this phenomenon requires a combined approach, including careful material selection, control of manufacturing processes, and the use of dehydrogenation techniques. HE can be mitigated through surface treatments that block hydrogen ingress and by optimizing the microstructure and composition of the material, focusing on alloying elements and phases that reduce hydrogen diffusion and accumulation.

With the increasing use of hydrogen as an energy source, the study and control of this problem are becoming increasingly relevant, especially in the production and storage of hydrogen in electrolyzers and high-pressure tanks.

4.5 Risk Studies for the Safety of Green Hydrogen Plants

Large-scale water electrolysis industrial plants use hydrogen and oxygen within the same equipment, separated by a membrane or diaphragm. Ensuring the safety of the process is essential; thus, risk studies are crucial for the safety of green hydrogen production via electrolyzers, as they identify and mitigate hazards such as hydrogen leaks (highly flammable), electrical risks, and the generation of pure oxygen. These analyses ensure the protection of workers, communities, and the environment, as well as ensuring compliance with standards and regulations. By evaluating failure scenarios and integrating safety measures from the initial design, risk studies reduce costs, optimize operations, and promote confidence in green hydrogen as a sustainable and safe energy solution, essential for the global energy transition. In this work, we will highlight a Hazard and Operability Study (HAZOP) analysis and initiating events for conducting a BOWTIE.

4.5.1 HAZOP

A HAZOP study, according to Crawley [26], is a structured analysis of a system, process, or operation for which detailed design information is available, to be carried out by a multidisciplinary team. The team proceeds by analyzing line by line or stage by stage a design of the process or operation. Although systematic and rigorous, the analysis also aims to be open and creative. This is done using a set of guide words in combination with system parameters to seek significant deviations from the design philosophy. A significant deviation is one that is physically possible—for example, no flow, high pressure, or reverse reaction.

In the technique, each step of the process uses guide words such as "none," "greater," "less," "inverse," and "part of" to predict possible failures. From this analysis, measures are proposed to eliminate or minimize the identified risks.

HAZOP is widely used in the industry to ensure safety and compliance with regulatory standards, being an essential tool for risk management in industrial processes. It allows organizations to develop proactive safety measures and assess possible deviations that may affect operations or personnel safety.

In Table 4, there is a HAZOP study conducted by [10] on a hydrogen production system EGA-9000, featuring an alkaline electrolyzer (Claind) with a hydrogen generation capacity of 1 Nm^3/h, a hydrogen storage system in the form of metal hydrides with a storage capacity of 5 kg of hydrogen, and a 5 kW PEM fuel cell to convert chemical energy into electricity when needed.

Table 4 HAZOP nodes for electrolyzers

	H_2 Storage system				
Deviation	Causes	Consequences	P S R	Salvagers	Recommences
Increase in Pressure (> 30 bar)	– The electrolyzer operated above its nominal pressure – High inlet flow with an increase in temperature – Improper maneuver: when opening or closing the cylinder valves or failures in entering the parameters of the Energy Management System	– Hydrogen loss – Formation of an explosive atmosphere and combustion, fire, explosion, deflagration	2 4 8	– Pressure sensor – Structure built with stainless steel materials (piping, valves, regulators, etc.) – PSV and Rupture Disk (RD) (opening pressure between 28 to 30 bar) – Two check valves	– Install two hydrogen sensors with alarm – Install emergency shutdown system – Procedure to close all manual valves in case of issues in the PSV and RD and the purification system – Procedure to purge the hydrogen storage system line
	Alkaline electrolyzer				
Deviation	Causes	Consequences	P S R	Salvagers	Recommences
– Low gas pressure (H_2)	– H_2 Loss	– Insufficient H_2 supply – Fire or explosion	2 4 8	– Pressure control – Lack of power in the electrolyze – Energy management system	– Install at least two more H_2 sensors

Source Adapted from Hadef et al. [10]

A risk matrix is used to evaluate the tolerability of a scenario based on the severity and frequency of undesirable events. The risk matrix used in the work of [10] is presented in Table 5. This matrix has four categories of likelihood of occurrence and five categories of severity; the combination of these categories results in three levels of risk: low, medium, and high.

In both scenarios presented, the risk was categorized as "High" (Remote with Severe Losses), which, according to the authors' classification, is an unacceptable risk. Therefore, recommendations for risk reduction were proposed after implementation, with the authors concluding that the EGA-9000 system is not a safe system.

Table 5 Risk matrix

Níveis de Severidade das Consequências (S)	Níveis de Probabilidade (P)			
	1—Improbable	2—Remote	3—Occasional	4—Likely
1—Minor damage	Low	Low	Low	Low
2—Damage	Low	Low	Medium	High
3—Severe damage	Medium	Medium	High	High
4—Severe Losses	Medium	High	High	High
5—Catastrophic	High	High	High	High

Source Adapted from [10]

The HAZOP study exemplified highlights the importance of this type of study for hydrogen production systems through electrolysis. The authors propose conducting a quantitative risk assessment through the modeling and simulation of fire and explosion risks, using appropriate tools (software and modeling model) to adequately quantify the risks.

4.5.2 Bow Tie

The Bow Tie method is a risk assessment technique used to analyze and demonstrate causal relationships in high-risk scenarios. The name derives from the shape of the diagram, which resembles a bow tie. This diagram provides a visual representation that summarizes all plausible accidental scenarios related to the hazard in question, also identifying the necessary control measures to manage these risks. Thus, the method clearly shows what needs to be done to effectively control risk scenarios [27].

A Bow Tie study conducted by the Institute for Sustainable Process Technology [28] on safety in low-carbon hydrogen production at an industrial scale, demonstrated in Table 6, found the following results for initiating events, which trigger a risk scenario, for containment loss scenarios in electrolyzers, along with their possible consequences, preventive measures, and mitigating measures:

4.6 Essential Tests for Electrolyzers

According to Tsotridis and [29], tests for hydrogen electrolyzers are essential to ensure their reliable and efficient operation. They include acceptance testing, which verifies the equipment's compliance with contractual specifications, and routine control testing, which ensures manufacturing quality. Other tests, such as conditioning and polarization curve, help optimize the performance and lifespan of the electrolyzer cells.

Main tests with electrolysers presented by Tsotridis and [29]:

install suitable sensors for safety monitoring of the installation to prevent the possible ignition of leaks and additional damage.

Regarding applicable standards, hydrogen detectors must comply with the manufacturer's risk assessment and must comply with IEC 60079-29-1 and/or ISO 26142. The detection system must initiate ventilation inside the hydrogen generating cabinets at an appropriate volume fraction of less than 1% of the hydrogen [19].

Hydrogen detection is essential in sectors such as chemical, petrochemical, metallurgy, and energy, ensuring greater safety and efficiency in industrial processes. Several sensor technologies are used to detect hydrogen, each with specific advantages and applications. Among the most common, catalytic, electrochemical, optical, and resistance-based sensors stand out. Materials such as palladium and advanced techniques, such as nanoparticles and acoustic waves, have been studied to improve the sensitivity and response time of sensors. The review identified several sensor technologies without indicating the best solution, as each one has disadvantages [30].

Collina et al. [30], presented several sensor technologies for hydrogen detection, highlighting their characteristics and applications. Catalytic sensors operate through the exothermic reaction between hydrogen and oxygen, with versions such as Pellistor and thermoelectric. Electrochemical sensors detect hydrogen by reactions that alter electrical properties, including amperometric and potentiometric types. Resistance-based sensors, on the other hand, use semiconductor metal oxides (MOx), whose resistance varies in the presence of hydrogen. Optical sensors measure changes in the optical properties of hydrogen-absorbing materials such as palladium (Pd), while sensors based on palladium nanoparticles (Pd NPs) take advantage of the high absorption capacity of this material, increasing the interaction surface. Finally, acoustic sensors use acoustic waves in piezoelectric materials, such as surface acoustic wave sensors.

According to Bozek and Nguyen [21], attention should be paid to infrared detection sensors, as they cannot detect hydrogen. In addition, gas detectors must be monitored, maintained, and calibrated periodically. A poorly maintained gas detector can pose an explosion hazard by providing a false sense of security. For this option to be viable, the gas detector must operate according to the manufacturer's specifications.

4.8 Storage

Hydrogen, being less dense than natural gas, presents significant challenges for storage and transportation, requiring more space for the same amount of energy. It is currently stored as a gas or liquid in high-pressure tanks for small-scale applications, but options such as compression in underground salt caverns, liquefaction at low temperatures, or conversion to liquid ammonia are explored for large-scale use. However, these processes require a large amount of energy, impacting costs. In addition, transporting hydrogen over long distances is complicated due to its low energy density, and its blending in existing pipelines faces technical challenges such as embrittlement, corrosion, combustion risks, and residual greenhouse gas emissions [31].

According to Calabrese et al. [11], there are several methods of storing hydrogen, including storage as compressed gas, liquid hydrogen, and solid-state storage, as described in Table 7. Gaseous hydrogen can be compressed in high-pressure tanks (350–750 bar), while liquid storage requires cooling to extremely low temperatures (– 253 °C) in insulated tanks. An alternative is cryocompressed storage (CcH_2), which combines cryogenic temperatures with moderate pressure (250–350 atm). In solid-state storage, hydrogen is stored physically or chemically in materials such as carbon nanotubes or metal hydrides, while liquid organic carriers involve hydrogenation and dehydrogenation cycles, which require heat to release the hydrogen. These methods are considered safe and practical, as they avoid the problems associated with storing hydrogen gas. However, many of the materials used are flammable and/or toxic, which can exacerbate the risks, especially due to the synergistic effects with hydrogen during release. These technologies combine harsh operating conditions (high temperature, high pressure, or flammable solid particles) with hydrogen, an intrinsically hazardous gas, resulting in high-risk systems.

Table 7 Main safety issues in hydrogen storage

Storage type	Security issues
Compressed hydrogen	High pressure, strong interaction with materials, high frequency of leakage
Cryogenic hydrogen	Low temperature, strong interaction with materials, difficult thermal management, risk of freezing, complex phenomena close to the release point
Cryo-compressed hydrogen	High pressure, low temperature, interaction with materials, difficult thermal management, freezing hazards, and complex phenomena near the point of release
Organic Liquid Hydrogen Carriers (LOHC)	Formic acid: corrosive, can cause severe burns to the skin and eyes DBT (Dibenzo-thiophene): low flammability, uncertain toxicity due to the presence of different regiosomers
Chemical storage	Methanol: toxic (ingestion of 56.2 g can be lethal; inhalation between 4000–13,000 ppm) Ammonia: toxic (lethal doses from 2700 ppm in 10 min, severe irritation from 220 ppm, lethal after 8 h at 390 ppm) and flammable

Source Adapted from Calabrese et al. [11]

Hydrogen storage presents significant challenges due to its physical and chemical properties, such as low energy density and the need for extreme operating conditions, such as high pressure or cryogenic temperatures. Methods such as compression in high-pressure tanks, liquefaction, cryocompressed and solid-state storage (metal hydrides or carbon nanotubes) offer viable solutions, but each brings safety issues such as leakage risks, toxicity, flammability and complexity in thermal management. In addition, the use of liquid organic carriers and ammonia requires chemical processes that require energy and special care. In order for hydrogen to be stored safely and efficiently on a large scale, it is essential to advance the development of more robust technologies, innovative materials, and regulations that minimize the associated risks, ensuring its viability as an integral part of the global energy transition.

4.9 Leaking Pipes—Inspections

The frequency of pipeline hydrogen leak inspections depends on a number of factors, including the frequency of hydrogen operations, local regulations, and risk assessments. While annual surveys may be suitable for some facilities, others may require more frequent inspections due to higher risk levels. Ultimately, the goal is to promptly identify and mitigate potential leaks to prevent accidents, environmental hazards, and economic losses.

Due to the physicochemical properties of hydrogen, such as its low molecular mass and high diffusion capacity, this gas is more prone to leaks compared to other fuels, requiring rigorous periodic inspections with specialized detectors to identify minimum concentrations. The frequency of checks should vary between 1 and 4 times a year, considering criteria such as the population density of the region and national technical regulations. Priority attention should be directed to critical components, such as flanged connections, valve stems, and compression connections, which, because they are subject to mechanical wear, thermal cycling, and pressurization, represent high-risk zones. The implementation of these preventive protocols aims to ensure the integrity of the pipes and mitigate potential failures, ensuring operational and public safety [32].

4.9.1 Hydrogen Pipeline Leak Inspection

The ASME B31.12-2023 standard that establishes technical requirements for the design of hydrogen piping and pipelines, with an emphasis on leak prevention and selection of materials compatible with the properties of the gas, presents some inspection and testing methods for detecting hydrogen leaks that can be used, as applicable, alone or in combination, according to documented procedures, indicating that other inspection and testing methods may be employed, provided that they are deemed appropriate and conducted in accordance with tried and tested procedures [33].

(a) **Surface inspection of hydrogen gas**

If there is a buried hydrogen pipeline or system, it is recommended to sample the atmosphere no more than 51 mm (2 in) above the ground surface whenever possible.

(b) **Subsurface Inspection with Gas Detector**

It is necessary to sample the atmosphere from the subsurface using a gas detector or other device capable of identifying 0.5% hydrogen in the air at the sampling point. The detector should be used in accessible openings, such as confined spaces, small substructures, inspection holes, and areas adjacent to the hydrogen facility.

(c) **Pressure drop test**

This test allows you to check for leaks causing pressure loss in an isolated segment of the hydrogen pipeline or network.

(d) **Ultrasonic Leak Test**

Leaks in exposed pipes can be detected with an instrument that identifies the ultrasonic energy generated by the gas escape.

(e) **Vegetation inspection**

In this inspection, it is possible to visually identify anomalies in the vegetation (such as dead or burned plants) that indicate leaks in the pipe. **Verify the causes**: perform visual checks of dead or burned vegetation.

(f) **Bubble testing**

Leaks can be detected by applying soapy water or other bubble-forming solutions to exposed pipes.

4.9.2 Risk Analysis for Hydrogen Pipelines

According to the European Industrial Gases Association [34], risk management in pipeline systems requires detailed safety analyses, especially in populated areas. These analyses assess environmental and operational impacts, identifying measures to reduce the likelihood and severity of failures. The level of detail considers population density, severity of incidents, and probability of occurrence. Individual **risk**—calculated as the chance of a fatality at a given location due to pipeline events—must be within regulatory limits. Local standards often establish acceptable criteria and limits, ensuring safety during the life cycle of the pipeline. A typical example of a thorough risk assessment for densely populated areas might include the following steps, as shown in Table 8.

Table 8 Example of risk analysis for H_2 leakage in pipelines

Dangers	Cause	Possible effects	Safeguards
Big/small H_2 leakage	– Damage by third parties – Use of incompatible materials and equipment in all operating conditions hydrogen embrittlement – External corrosion due to inadequate cathodic protection – Valves in the valves and valves – Over-pressurization of the pipe – Inadequate inerting procedure – Improper operation and maintenance of the pipeline – Abnormal loads due to landslides, floods, road crossings – Influence of other structures, such as high-power power lines – Damage due to an abnormal event in a parallel pipe – Road accident or fire in the vicinity of parts of the above-ground pipeline – Radiation from a vent fire or an explosion	– Release of the hydrogen, which ignites immediately or spontaneously, rises and disperses into the open air, possibly with delayed ignition In contact with ignition sources Possible occurrence of: – Jet *fire* – Cloud fire (*flashfire*) – Explosion	– Control of third-party interference – Increased tube thickness – Non-destructive testing on welds – Inerting of the pipe – Pipeline marking – Isolation valves – Excess flow or low pressure shut-off valves – Mass balance leak detection – Aprofundar o pipeline – Physical protections: concrete cladding or cladding, concrete slabs – Rerouting of pipelines – Reduction of piping operating pressure – Operational procedures, including: inspection programs, corrosion control programs, emergency planning, staff training, third-party information, collaboration with local authorities

Source Adapted from EIGA [34]

5 Results and Discussion

The study showed that the production of green hydrogen via electrolysis, although essential for the energy transition, faces technical and safety challenges that require integrated approaches. The comparison between alkaline and PEM electrolyzers revealed significant trade-offs: while alkaline systems, mature and with a lower initial cost (USD 500–1400/kW), are suitable for large-scale plants, their intrinsic corrosivity (due to the use of KOH) increases maintenance costs and degradation risks. PEM electrolyzers, on the other hand, with high power density (< 4.4 mW/cm^2) and the ability to operate under high pressures (up to 70 bar), are ideal for applications that require flexibility, but face limitations such as accelerated degradation (20,000–60,000 h) and dependence on noble metals (platinum, iridium), increasing initial investments.

The hazardous area classification highlighted critical normative divergences. While ABNT NBR IEC 60079-10-1 defines risk zones with "negligible" extension, Petrobras Standard N-2918 adopts more conservative criteria, delimiting fixed areas of 4.5 m (horizontal) and 7.5 m (vertical) for zones 1 and 2. This approach reflects practical concerns with hydrogen's high flammability (LIE 4–75%) and its low ignition energy (0.02 mJ), reinforcing the need for mechanical ventilation and distance between equipment in semi-confined environments.

HE has emerged as a central challenge in materials exposed to pressures greater than 30 bar. The penetration of hydrogen atoms into metallic microstructures, such as carbon steel and nickel alloys, reduces ductility and mechanical strength, increasing the risk of catastrophic failures. Mitigation strategies include surface treatments (nitriding, aluminum oxide coatings) and modifications in the composition of alloys, with the addition of elements such as molybdenum to stabilize carbides and increase the fraction of austenite, a phase less susceptible to HE.

Risk analyses, such as HAZOP and *Bow Tie*, have proven to be indispensable tools for identifying critical scenarios. In a pilot alkaline system, HAZOP classified as "unacceptable" risks such as excessive pressures (> 30 bar) and leaks, recommending redundant sensors and relief valves (PSV). *Bow Tie, on the other hand*, highlighted the importance of preventive measures (specification of resistant materials) and mitigating measures (forced ventilation, fire barriers), in line with ISO 22734, which requires emergency stops for H_2 concentrations above 2%.

Hydrogen storage and transportation presented additional complexities. Methods such as compression (350–750 bar) and liquefaction (–253 °C) involve risks of leakage and energy costs, while alternatives such as metal hydrides and organic carriers (e.g., ammonia) face challenges of toxicity and secondary combustion. Periodic inspections, as per ASME B31.12-2023, and protocols such as ultrasound and pressure drop tests have reduced incidents by 40% in pilot units, validating the importance of predictive maintenance.

In summary, the safe production of low-carbon hydrogen requires integration between advanced technologies, strict regulations, and proactive risk management. The combination of quantitative analysis, resilient materials, and systematic inspections is crucial to balance efficiency, costs, and safety, enabling hydrogen as a pillar of global decarbonization.

6 Conclusion

The production of low-carbon hydrogen via electrolysis emerges as a promising strategy for the energy transition and reduction of carbon emissions. However, the process brings with it significant industrial risks, related to the flammability of hydrogen, the corrosion of materials and the complexity of storage systems. Such risks can be minimized by adopting advanced monitoring technologies, rigorous selection of materials, and implementing a robust personnel training program.

This study demonstrated that, although technology is feasible, its implementation on an industrial scale requires overcoming technical, operational and security challenges. The comparative analysis between alkaline electrolyzers and PEM showed that the choice of technology depends on contextual factors: alkaline systems are economically advantageous for large projects, while PEM systems stand out in applications that demand high pressure and operational flexibility, even with high initial investment costs.

Safety emerges as a critical pillar, especially given hydrogen's intrinsic properties, such as high flammability (4–75% LIE) and low ignition energy (0.02 mJ). The adoption of strict regulations, such as **Petrobras Standard N-2918**, proved to be essential to mitigate risks in semi-confined environments, with clear guidelines for classification of areas and ventilation. In addition, phenomena such as hydrogen embrittlement (HE) require careful selection of materials (e.g., austenitic steels) and mitigation strategies, such as surface treatments and microstructural modifications, to ensure the integrity of components under pressure.

Risk management tools, such as **HAZOP** and ***Bow Tie***, have proven effective in proactively identifying critical scenarios such as leaks and over-pressurization, guiding the implementation of safeguards such as redundant sensors, relief valves (PSV), and automatic depressurization systems. Hydrogen storage and transportation, on the other hand, require innovative solutions, such as metal hydrides and carbon nanotubes, to balance safety, costs, and energy efficiency.

To make green hydrogen viable as an energy carrier, it is essential to integrate technological advances, adaptive regulations, and multi-sectoral cooperation. Cost reduction, through economies of scale and innovation in materials (e.g., platinum replacement in PEM catalysts), combined with the training of professionals in risk management, will be decisive to consolidate this alternative. In the Brazilian context, the synergistic potential between abundant renewable resources (wind, solar) and industrial *expertise* positions the country as a relevant player in this emerging market.

In summary, the success of the green hydrogen economy depends not only on technical advances, but on an ecosystem that prioritizes safety, sustainability, and operational resilience, ensuring its effective contribution to a decarbonized energy future. Security in the production of green hydrogen requires a balance between the energy potential of hydrogen and the proactive management of its intrinsic risks. Investments in research, development of adaptive regulations, and the integration of best industrial practices are essential steps to ensure that this promising technology contributes effectively and sustainably to the decarbonization of the global economy.

References

1. de Assis TF, Monteiro TGM, de Abreu VHS, D'Agosto MDA, Santos AS (2022a) Enabling the green bonds market for sustainable transport projects based on the measure/monitoring, reporting and verification method. In: Carbon footprints of manufacturing and transportation industries. Springer, Singapore, pp 1–24
2. de Assis TF, Ricci LM, Monteiro TGM, de Abreu VHS, D'Agosto MDA, Santos AS (2022b) Sustainable transport indicators and mitigation actions applied to the green bond principles. In: Carbon footprints of manufacturing and transportation industries. Springer, Singapore, pp 139–169
3. de Abreu VHS, Pereira VGF, Proença LFC, Toniolo FS, Santos AS (2023) A systematic study on techno-economic evaluation of hydrogen production. Energies 16(18):6542
4. Molkov V (2012) Hydrogen safety engineering: the state-of-the-art and future progress. In: Sayigh A (ed) Comprehensive renewable energy, vol 4. Elsevier, pp 77–109. https://doi.org/10.1016/B978-0-08-087872-0.00418-2
5. Ball M, Weeda M (2015) The hydrogen economy—vision or reality? Int J Hydrogen Energy 40(25):7903–7919
6. De Abreu VHS, Coralli A, Proença LFC, Santos AS (2023b) Avaliação Do Ciclo de Vida Como Ferramenta Para Medir a Eficiência Ambiental Do Hidrogênio Renovável [Life Cycle Assessment as a Tool for Measuring the Environmental Efficiency of Renewable Hydrogen]. Impact Proj. Impact Proj
7. de Abreu VHS, Da Ros Hollanda DL, Proença LFC, Bahiense L, Santos AS (2024) A systematic review on renewable hydrogen application in the land transportation sector. Intell Transp Syst Adv Technol 9–31. https://doi.org/10.1007/978-981-97-0515-3_2
8. de Abreu VHS, Santos AS (2024) Promoting a sustainable energy footprint: Brazilian initiatives for renewable hydrogen. In: Energy footprint and sustainability: case studies. Springer, Cham, pp 23–36. https://doi.org/10.1007/978-3-031-63057-6_3
9. Associação Brasileira de Normas Técnicas—ABNT (2022) NBR IEC 60079-10-1: Atmosferas explosivas – Parte 10–1: Classificação de áreas – Atmosferas explosivas de gás. 3 ed. Rio de Janeiro
10. Hadef H, Negrou B, Ayuso TG, Djebabra M, Ramadan M (2020) Preliminary hazard identification for risk assessment on a complex system for hydrogen production. Int J Hydrog Energy [S.l.] 45(15):8876–8890. Disponível em: https://www.sciencedirect.com/science/article/abs/pii/S0360319919340091?via%3Dihub. Acesso em: 10 out. 2024
11. Calabrese M, Portarapillo M, Di Nardo A, Venezia V, Turco M, Luciani G, Di Benedetto A (2024) Hydrogen safety challenges: a comprehensive review on production, storage, transport, utilization, and CFD-based consequence and risk assessment. Energies 17:1350. https://doi.org/10.3390/en17061350
12. Valdés R, Rodríguez LR, Lucio JH (2012) Procedure for optimal design of hydrogen production plants with reserve storage and a stand-alone photovoltaic power system. Int J Hydrogen Energy 37:4018–4025
13. Ursúa A, Gandía LM, Sanchis P (2012) Hydrogen production from water electrolysis: current status and future trends. Proc IEEE 100(2):410–426
14. Ratib MK et al (2024) Electrical circuit modeling of proton exchange membrane electrolyzer: the state-of-the-art, current challenges, and recommendations. Int J Hydrog Energy 49(pt. C)
15. Hancke R, Bujlo P, Holm T, Ulleberg Ø (2024) High-pressure PEM water electrolyser performance up to 180 bar differential pressure. J Power Sources [S.l.], v 601
16. International Renewable Energy Agency (IRENA) (2020) Green hydrogen cost reduction: scaling up electrolysers to meet the 1.5°C climate goal. IRENA, Abu Dhabi. Disponível em: https://www.irena.org/-/media/Files/IRENA/Agency/Publication/2020/Dec/IRENA_Green_hydrogen_cost_2020.pdf. Acesso em: 25 jan. 2025
17. Michigan Technological University (MTU) (2024) Process safety. Disponível em: https://www.mtu.edu/chemical/facilities/safety/process/. Acesso em: 28 dez. 2024

18. Pascuzzi S et al (2016) Hazards assessment and technical actions due to the production of pressured hydrogen within a pilot photovoltaic-electrolyser-fuel cell power system for agricultural equipment. J Agricult Eng 47(2):88–93. https://doi.org/10.4081/jae.2016.507
19. ABNT (Associação Brasileira de Normas Técnicas) (2024) ISO 22734: Geradores de hidrogênio por eletrólise da água – aplicações industriais, comerciais e residenciais. 1. ed. Rio de Janeiro: ABNT
20. Kleszcz S, Assadi M (2023) Hydrogen safety considerations: mitigating risks and securing operations in enclosed spaces. In: IOP conference series: materials science and engineering, vol 1294, p 012057. Disponível em: https://www.researchgate.net/publication/376747562_Hydrogen_safety_considerations_Mitigating_risks_and_securing_operations_in_enclosed_spaces. Acesso em: 9 jan. 2025
21. Cekerevac Z, Cekerevac D (2022) Considerations for hazardous area classification of modular hydrogen facilities. HazardEx 29 abr. 2022. Disponível em: https://www.hazardexonthenet.net/article/190564/Considerations-for-hazardous-area-classification-of-modular-hydrogen-facilities.aspx. Acesso em: 17 fev. 2025
22. Associação Brasileira de Normas Técnicas—ABNT (2022) NBR IEC 60079-10-1: Atmosferas explosivas – Parte 10–1: Classificação de áreas – Atmosferas explosivas de gás. 3 ed. Rio de Janeiro
23. PETROBRAS (2020) N-2918: Atmosferas Explosivas – Classificação de Áreas. Rio de Janeiro
24. Rigas F, Amyotte P (2013) Hydrogen safety. CRC Press, Boca Raton
25. Li X et al (2020) Review of hydrogen embrittlement in metals: hydrogen diffusion, hydrogen characterization, hydrogen embrittlement mechanism and prevention. J Mater Sci Technol 62:96–117. Disponível em: https://link.springer.com/article/, https://doi.org/10.1007/s12540-020-00658-9. Acesso em: 10 jan. 2025
26. Frank C, Brian T, Edward P (2015) HAZOP: Guide to Best Practice, 3rd edn. Elsevier, Amsterdam
27. Sankar V et al (2016) Application of Bow-Tie analysis in risk management. IJSRD Int J Sci Res Dev 4(9):2321–0613
28. Institute for Sustainable Process Technology (ISPT) (2023) Safety aspects of green hydrogen production on industrial scale. Hydrohub Innovation Program. Disponível em: https://www.h2-mobile.fr/uploads/doc_20231114192000.pdf. Acesso em: 23 jan. 2025
29. Tsotridis G, Pilenga A (2018) EU harmonised terminology for low temperature water electrolysis for energy storage applications. Joint Research Centre (JRC), The European Commission's Science and Knowledge Service
30. Collina G, Bucelli M, Paltrinieri N (2024) Multi-stage monitoring of hydrogen systems for improved maintenance approaches: an extensive review. Int J Hydrog Energy 105:458–480. Disponível em: https://doi.org/10.1016/j.ijhydene.2025.01.112. Acesso em: 27 dez. 2024
31. Energypedia (2023) Green hydrogen. Disponível em: https://energypedia.info/wiki/Green_Hydrogen. Acesso em: 20 out. 2024
32. IGC (International Gas Carrier) (2014) Hydrogen pipeline systems. Doc 121/14. Revision of Doc 121/04. [S.l.]: IGC
33. ASME (American Society of Mechanical Engineers) (2023) ASME B31.12-2023: Hydrogen piping and pipelines—ASME code for pressure piping, B31. ASME, New York
34. European Industrial Gases Association (2014) Hydrogen pipeline systems. IGC Doc 121/14. Revision of Doc 121/04. EIGA, Brussels
35. SENAI (2024) Classificação de área para atmosferas explosivas da planta piloto de eletrólise para estudo da valoração da cadeia de hidrogênio de baixo carbono. Organização de Juan Alberto Chávez Ruiz; Amora Vieira Cavalcante. Natal
36. Rashidi S, Mostofi SN (2021) Review on the safety challenges of hydrogen production, storage, and utilization. J Hydrog Energy 46(11):7910–7929

Integral Use of Biotechnology in Textiles-An Innovative Way to Reduce Carbon Footprint

R. Surjit, P. Yazhini, M. Amirthavarshini, R. Rithika Sri, and B. P. Pandiyammal Devi

Abstract Industrial operations lead to significant greenhouse gas emissions, with carbon dioxide being the major threat. The industrial carbon footprint is the measure of the amount of carbon dioxide released during each process of industrial production. The textile and apparel industry emits a very large amount of carbon through its processes, starting from fibre production to disposal. It has been a perennial problem, and measures are being taken to minimise the carbon footprint. One of the measures taken is the use of biotechnology. Biotechnology has been used at a great level in the textile and apparel industry as it unleashes a plethora of opportunities to reduce the carbon footprint. Biotechnology provides a variety of creative options for lowering industrial carbon footprints, such as carbon capture and utilisation of bioenergy and biofuels. Biotechnology also allows us to create more sustainable related production processes, which reduce the energy usage and greenhouse gas emissions. It can also be utilized to increase the efficiency of industrial processes, hence lowering the wastage and the pollution caused by these activities. This chapter discusses the various biotechnology-based strategies used to minimise the industrial carbon emissions generated by the textile and apparel industry. The application of biotechnology in the areas of fibres and fabrics production, textile finishing using enzymes, sustainable dyeing techniques, and bio-based recycling of textile wastes is discussed in this chapter. The chapter also divulges information on circular economy solutions through biotechnology usage to reduce industrial carbon footprint. This chapter also gives an overview of the current biotechnological measures taken in the textile and apparel industry for lowering the industrial carbon footprint and discusses the future possibilities of their usage.

Keywords Sustainability · Biotechnology · Finishing · Greenhouse gases · Biodegradability

R. Surjit (✉) · P. Yazhini · M. Amirthavarshini · R. Rithika Sri · B. P. Pandiyammal Devi
Department of Fashion Technology, PSG College of Technology, Coimbatore, Tamil Nadu, India
e-mail: rst.fashion@psgtech.ac.in

S. S. Muthu (ed.), *Reduction of Industrial Carbon Footprint*, Environmental Footprints and Eco-design of Products and Processes,
https://doi.org/10.1007/978-3-032-10913-2_6

1 Introduction

Biotechnology is the process that utilizes biological systems like enzymes, microorganisms, and plant-based materials to create sustainable and innovative fabrics. Biotechnology has been rapidly evolving for the past few years. There is a biodegradable option for many products right now, where the microorganisms are used to break down the long and complicated chains to decompose the product to obtain a greener environment. Many new inventions use natural dyes more effectively and durably. These inventions allow us to overcome many issues faced in the industry. The biotechnology also helps in the creation of new processes in fiber production by increasing its strength, stability, and durability for an extended period. Moreover, the rapid growth of the enzymatic approach in the textile processes revolutionized the apparel industry by making the processes cheaper, safer, and healthier. Most of the traditional textile industries have started to take a more enzymatic approach to processes like desizing, scouring, and bleaching. Sustainability plays a key role in achieving all the measures and processes mentioned above. Sustainability merges ideas and technologies to create many products while conserving resources for future use and limiting the amount of pollution emitted. This chapter discusses the various biotechnology-based strategies used to minimise the industrial carbon emissions generated by the textile and apparel industry. The application of biotechnology in the areas of fibres and fabrics production, textile finishing using enzymes, sustainable dyeing techniques, and bio-based recycling of textile wastes is discussed in this chapter. The chapter also divulges information on circular economy solutions through biotechnology usage to reduce industrial carbon footprint. This chapter also gives an overview of the current biotechnological measures taken in the textile and apparel industry for lowering the industrial carbon footprint and discusses the future possibilities of their usage.

2 Sustainability Measures to Reduce Carbon Footprint

Sustainability is the process of satisfying the present needs without compromising the resources for future generations. It ensures that future generations also enjoy a good lifetime and a healthy planet. To sustain a pollution-free and healthier environment, we need sustainability. Sustainability has been a growing trend for the past few years in the textile industry. There are many ways to induce sustainability, such as the usage of eco-friendly materials, bringing up the concept of the circular economy, and much more. Organic cotton, linen, bamboo, and recycled fibres are some of the eco-friendly materials that can be used to create a sustainable garment. The circular economy is also a great way to attain sustainability. The materials are reused and recycled rather than being discarded.

2.1 *Circular Economy in Textiles*

The textile industry is one of the most polluting industries since a large amount of waste is produced in each process in manufacturing a product. These substances release a huge amount of toxins into the air, water, and other natural bodies. So, to sustain a healthier environment, we use the circular economy. The main objective of the circular economy of textiles is to minimize the amount of waste produced by reducing, reusing, or recycling the waste back into the manufacturing process, which maximizes the lifespan and quality of the product. Recycling is the first process that comes to mind when thinking about the circular economy. When textiles are recycled, each part of the textile product is broken down to create a new product. However, the main issue is that the textile may contain trims, such as buttons or zips, which make it difficult to recycle the product. These items fall into the reuse category rather than recycling, as they have a longer lifespan than textile materials. There are also some new innovative methods to overcome this difficulty, such as.

- **Chemical recycling**: In chemical recycling, the textile materials are treated with chemicals to break down the fiber to produce alternate, higher-quality products.
- **Mechanical recycling**: In mechanical recycling, the textile materials are torn into pieces, which are processed to produce new fibers or garments.

Next comes the reduction method. There are plenty of ways to minimize the carbon footprint through the reduction methods. Some of the methods are to use low and ultralow liquor ratio machines, which reduce the water consumption and simultaneously the energy consumption, or by combining the scouring and the bleaching, which also reduces the water consumption by combining two baths in one, or by using enzymes for various process which will be father discussed later in this chapter. The enzymes reduce the number of processing steps by combining them, which ultimately reduces the amount of waste produced. Finally, there is the reuse. By reusing a material, the use of resources, manpower, and energy is preserved.

Even though the circular economy reduces the overall wastage in the manufacturing processes of a product, the eco-friendly dyeing and finishing processes reduce the amount of chemicals used in these processes and also the amount of toxins released into the environment.

2.2 *Eco-friendly Finishing*

The eco-friendly finishing processes reduce the environmental impact produced by the manufacturing processes, as the traditional finishing processes release a lot of harsh chemicals and consume a lot of energy and water. This increases the pollution and the carbon footprint in the environment. There have been many technologies that have been invented to help us achieve eco-friendly finishing. Some of them are.

- **Plasma technology**: The plasma processing can reduce the amount of carbon emitted with the help of renewable electricity. This application involves activating stable small molecules such as CO_2, CH_4, and N_2, which can outperform their thermocatalytic capacity and ultimately reduce the carbon emission.
- **Nanotechnology**: The nanotechnology doesn't require as much raw material than the other technologies which reduces the carbon footprint as the amount of processing depends on the amount of the raw materials required. The nano technology doesn't require higher temperature during its process which also reduces the footprint.
- **Ultrasonic finishing**: In ultrasonic finishing, the amount of materials used is reduced, resulting in a decreased carbon footprint.

All the above-mentioned finishes help to reduce the carbon footprint in a lot of different ways, but there are also some other processes which help in the control of the pollution caused by the manufacturing processes by various procedures.

2.3 Bio-tech-based Water Conservation

The disposal of harmful chemicals from the industries contaminates the water, which causes a huge risk to the environment. This contamination helps in the growth of harmful microbes, which can destroy the health of the environment. Some microalgae species like Scenedesmus, Chlorella, Phormidium, and Chlamydomonas remove the nutrients from the water, making it easier to contaminate, thus polluting the environment. The chromium and mercury produced by the dyeing industries are toxic to living things.

Some bioremedial agents, like microorganisms, can convert the toxic materials into water and CO_2, which can further be processed and degraded by the microbes, ultimately reducing the carbon footprint. The mechanism of degrading the pollution can be divided into two categories.

- **Mobilization**: It uses the enzymatic oxidation, biostimulation, bleaching, and the enzymatic reduction procedure to degrade the waste.
- **Immobilization**: It uses bioaccumulation, complexation, biosorption, and precipitation to degrade the waste.

The marine microorganism is one of the most superior in enzymatic production than plants or animals, as more than 70% of the world is full of marine microorganisms [25]. Research has shown that the development of Polyhydroxyalkanoates has been increased due to the marine bacteria. To control the harmful water bodies and to conserve the water for future use, the marine biotechnology processes are done in the aquatic regions. The organic matter present in the textile waste can be reduced in a low space by converting it to Methane and CO_2 by anaerobic digestion, and treats the waste containing 20,000 mg/L of COD with low cost, and also reduces the sludge formation. It saves time, water, and energy that is needed in a traditional method at an affordable price. Some of the examples are the combined bleach clean-up and dyeing, and combined desizing and biosourcing [32].

2.4 Sustainable Fibres

Sustainable fibres have been rapidly evolving for the past few years. These fibres don't harm the environment, unlike synthetic fibres like polyester and nylon. Cellulosic fibres are one such example of sustainable fibres that consume low energy and can also be easily biodegradable. The lignocellulosic fibres are usually sourced from plants and wood. It consists of hemicellulose, lignin, and cellulose, which are biodegradable and are used in various industries.

The natural fibres are usually divided into two types.

- **Wood**: The fibres that are produced from the wood or logs. E.g., Orchids.
- **Non-wood**: The fibres that are produced from plants and vegetable fibres. E.g., Lotus fibre.

The lignocellulose falls under the category of wood fibres, while lotus fibre, pineapple leaf, and banana fibres fall in the category of non-wood fibres. The wood fibres have special characters like tensile strength and dimensional stability. These fibres help the environment by being reusable and degradable. Since these require less energy, the amount of carbon footprint is reduced.

3 Biodegradable Textiles

Biodegradable fibers are referred to as the fibers that will decompose quite easily and naturally by the use of microorganisms like bacteria, fungi, etc. Biodegradable fibers are used in textiles, packaging, agriculture, and in biomedical. A major focus of biodegradable fibers is to reduce environmental waste and pollution.

3.1 Polylactic Acid (PLA) Textiles

Polylactic acid was first manufactured by Dow in 2002. Polylactic acid fiber is the only biobased and biodegradable polymer. Several aliphatic polyester fibers, including polycaprolactone (PCL), polyhydroxyalkanoates (PHA), and polybutylene succinate (PBS) fibers, are also biodegradable, but these fibers are made from unsustainable petrochemical sources.

Features of polylactic acid:

- More hydrophilic

- Good resilience
- Unaffected by UV light
- Controllable shrinkage
- Controllable thermal bonding temperatures
- Low flammability and smoke generation.

The use of polylactic acid reduces the carbon dioxide footprint comparison to other traditional petroleum-based plastics, because this fixes the significant amount of carbon dioxide. The main feature of PLA is its physical properties can be modified according to its purpose. PLA is relatively hard and its glass transition temperature is 60–70 °C while the melting at 170–180 °C [11].

3.2 Biodegradable Nonwovens

Non-wovens is one of the fastest-growing textile sectors. These non-wovens can be produced from both natural and synthetic fibers, directly from the polymers or biopolymers, by a variety of techniques like web creation and bonding. These non-woven materials are commonly used as single-use products or short-life products, which creates a lot of waste during their disposal. In recent years, the demand for the use of biodegradable materials has increased drastically, especially in the areas of disposable products like surgery masks, tissues, etc. [21]. The real challenge of the non-woven industry is the production of huge volumes of products, which leads to huge volumes of wastage, is the search for biodegradable polymers. The nonwovens are classified based on the web bonding:

- Thermal bonding:
  - o Sonic bonding
  - o Calendering
  - o Hot air pressing.
- Mechanical bonding:
  - o Needle punching
  - o Spun lead
  - o Stitch bonded.
- Chemical bonding:
  - o Foam coating
  - o Print bonding
  - o Impregnating
  - o Spraying.

3.3 *Biodegradable Textile Finishes*

The biodegradable textile finish is an alternative method for the traditional finish, which requires a huge amount of water and numerous chemicals. The traditional textile finishing method requires high energy. This leads to a significant effect on the environment. Therefore, biodegradable textile finishes are developed and used to minimise the chemical and energy usage and also for less amount of water usage. Trends with sustainable concepts, where natural materials such as minerals and plant extracts are used in biodegradable textile finishes, are now being used by a large number of people. It will not only protect the environment from pollution but also control climatic changes, carbon footprint, and be user-friendly [16]. Innovative textile characteristics include antibacterial, flame-retardant, fog-harvesting, and easy-to-clean or dirt-repellent capabilities [26]. In addition, there are textiles for electronics based on graphene, carbon nanotubes, and other nanomaterials; conductive textiles; textiles for sensors; textile-fixed catalysts; textiles as tissue engineering substrates; and textiles for O/W separation. All of these advancements have been made feasible by incorporating new methods to textile finishing, such as applicable surface modification techniques and applying natural biomimicry principles. There are also finishes made with the assistance of the microbes, which will be discussed later in this chapter. This biological approach reduces the carbon footprint by a huge amount.

The application of biodegradable finishes is classified as follows:

1. Pad-dry-cure—It is a popular textile finishing method for applying chemical dyes or treatments to textiles. It involves three steps, which are
   - o Pad—The textile will pass through a chemical solution like dye, softener, water repellent, or anti-wrinkle agent.
   - o Dry—Moisture gets removed from the padded fabric by drying it, usually with hot air. Even though it doesn't yet permanently cure the chemicals, this aids in getting the fabric ready for curing.
   - o Cure—Heat is applied to fix the chemical onto the fabric.
2. Exhaust—A method involving environmentally friendly dyeing or finishing in which the fabric progressively absorbs biodegradable chemicals or colors from a solution. Because the remaining bath contains non-toxic, compostable residues, the process ensures low waste production and sustainable disposal.
3. Foam finishing—It is a textile finishing technique that distributes finishing chemicals to the surface of fabric via foam instead of liquid. Air as well as a liquid chemical solution are combined to make the foam, which is stable, lightweight, and uniformly applicable.
4. Microencapsulation—Through the integration of microencapsulation technology, a substance can be contained within a microsphere, also known as a microcapsule, which is a tiny sphere with an average diameter ranging from 1 μm to several hundred micrometers.

5. Spray—Using spray nozzles or guns, the spray method distributes a thin mist of biodegradable finishing substances directly onto textile materials' surfaces.

3.4 End-of-Life Management for Biodegradable Textiles

In recent years, the principles of sustainability and the circular economy have gained popularity in the textile industry. Textiles should be manufactured in such a way that they return to the natural cycle at the end without creating further harm to the environment [14]. Natural or synthetic fibers that are broken down into small components by some microorganisms like bacteria, fungi, or algae, which turn them into natural components, including water, carbon dioxide, and biomass, are termed biodegradable textiles. The breakdown process occurs without generating any harmful residues or microplastics. The biodegradability of textiles is mostly determined by the amount of chemicals used in their life cycle. The more chemicals used, the longer it takes for the fabric to biodegrade, which causes harm to the environment. Several biodegradable textiles are selected based on their ecological consequences, rate of complete depletion, and degree of decomposition [27]. For biodegradable textiles, the following strategies are used to minimize their negative effects on the environment through their naturally occurring breakdown capabilities, such as,

1. Compositing: For textiles that require specific moisture, temperature, and bacteria growth to break down effectively, industrial compositing is a suitable one. And natural fibers such as cotton, linen, and hemp can be composted at home if it is not treated with artificial dyes or finishes.
2. Anaerobic digestion: Biodegradable textiles are also processed in anaerobic digester systems for generating biogas and nutrient-rich digestate.
3. Mechanical Recycling: Recycling through machines before the biodegradable textiles undergo biodegradation. Biodegradable textiles can be mechanically shredded and upcycled as low-quality products like stuffing, insulation, or industrial wipes.

4 Microbial-Based Textile Finishing

The microbial-based textile finishing involves treating fabrics with agents that inhibit or kill microorganisms and enhance hygiene and durability. This process protects the fabrics from damage caused by bacteria and fungi. The primary goals of antimicrobial finish are to protect textiles by preventing the damage caused by microbial growth such as mildew, rot and staining and enhance hygiene by reducing the risk of cross infection and odor development on textiles and also to improve the performance like maintaining cleanliness of textiles particularly those used in healthcare and extending lifespan of the fabrics. Plant extracts are attractive sources of the eco-friendly antimicrobial finish, for example: Neem (*Azadirachta indica*) of the Meliaceae family and

Papaya (*Carica papaya*) of the Caricaceae family, which is rich in vitamin A, C, and E antioxidants [3]. Since the microbes don't require much energy or temperature, the amount of carbon footprint is reduced.

4.1 Natural Finishing Agents

In recent years, there has been a passion towards natural herbs as antimicrobial agents. Eco-finishes involve substances that are obtained from flora and fauna. They have many advantages, like non-harmful, non-irritant, biodegradable, and economically easily available. Both clothing and textile materials are the main reasons for the growth of microorganisms. They act as carriers of some of the microorganisms. For example, pathogenic microorganisms, smell-producing bacteria, and fungi. In the present-day world, the vast majority of us have knowledge or awareness about our hygiene and cleanliness. Some plant extracts act as natural finishing agents. These agents are rich chemically and control a variety of different allergies, infections, and diseases. For example, *Aloe vera*, neem, tulsi leaf, amla, eucalyptus, clove, datura stramonium, calotropis, natural honey, and sericin from the silkworm. *Aloe vera* and neem plants have traditional and pharmacological properties. The natural finishing reduces the global carbon dioxide emissions drastically.

- Neem—Giving a neem finish to the garment will be effective against cough and breathing problems. Neem extracts have more than 300 active compounds like azadirachtin, salannin, nimbin.
- *Aloe vera*—*Aloe vera* finishes garments are safe for skin with excellent moisture-retaining properties.

4.2 Antimicrobial Finish

Antimicrobial finish is the use of antimicrobial agents that are used to prevent infections and diseases caused by pathogens. There are two main features for the antimicrobial finish:

1. Protecting the textile from the odor-causing bacteria.
2. Protecting the textile from the damage caused by the pathogens.

The growth of microorganisms on textiles leads to the creation of unpleasant odours, a reduction in the fabric's strength, and discoloration of the fabric. To prevent this, the textiles are treated with various antimicrobial agents to suppress or prevent textiles from degrading. The natural finishing agents will also give an antimicrobial finish. Antimicrobial finishing agents have been introduced to many natural and synthetic textiles by a simple and common method called the pad-dry-cure process. The pad-dry-cure method is one of the most commonly used traditional techniques

- Application:
  - o Organic UV absorbers:
    Bioactive substances with UV-absorbing properties are gaining recognition as sustainable and environmentally friendly alternatives for natural textile finishes. Various plant-based extracts, including those from marine marigold (*Tagetes erecta*), macroalgae (seaweeds), *Scutellaria baicalensis Georgi,* and *Aloe vera,* have been utilized in textile treatment due to their strong potential to block ultraviolet radiation.
- Inorganic UV absorbers:
  For inorganic UV blockers, current research trends emphasize the development of advanced textiles that combine UV protection with multiple functionalities, including antimicrobial effects, resistance to oil and water, easy-care performance, resistance to UV-induced degradation, electrical conductivity, flame retardancy, and thermal stability.

4.5 Flame Retardant Finish

Flame-retardant (FR) finishing is a crucial treatment essential for producing textiles that offer protection against fire hazards. Polymers are generally regarded as highly flammable materials, and their widespread use in households can inadvertently raise the overall fire load. Two primary parameters used to evaluate a material's flammability are the LOI (Limiting Oxygen Index) and THR (Total Heat Release). The minimum concentration of oxygen required to maintain combustion is known as LOI, while the THR indicates the amount of heat energy emitted during the material's burning process. The toxicity of fire gases is a crucial factor in assessing fire hazards. When polymers burn, they can emit harmful substances such as carbon monoxide, carbon dioxide, halogenated compounds, organic irritants, and soot. Inhalation of these toxic gases can cause immediate health effects in exposed individuals, including coughing, difficulty breathing, and dizziness. For instance, the combustion of polyvinyl chloride and Nylon 6.6 generates significant amounts of hazardous gases, such as hydrogen chloride and hydrogen cyanide, respectively. When polyurethane foam is exposed to temperatures above 300 °C, it decomposes and emits yellow smoke along with toxic compounds like isocyanates and amines [39]. The emission of these combustion byproducts into the atmosphere also contributes to deteriorating air quality, resulting in long-lasting environmental pollution and harm. A flame retardant is a substance incorporated into polymers to regulate their combustion by disrupting the physical and/or chemical reactions that occur during burning. These additives are typically classified based on how they function. Flame retardants can inhibit fire propagation through physical means, such as diluting flammable gases, absorbing heat (endothermic cooling), or creating a protective barrier using nanoparticles, or through chemical mechanisms like suppressing flame reactions, promoting char formation, causing swelling (intumescence), or facilitating melt dripping. Three common strategies can be employed to enhance the fire-resistant characteristics of polymeric materials: incorporating flame-retardant additives into polymer composites, applying flame-resistant surface coatings, and chemically modifying or functionalizing the polymer chains. The flammability of a material tends to increase with greater oxygen availability, which can accelerate fire development. However, in some cases, excessive ventilation may slow fire growth by dispersing heat away from the flame. This variability makes it challenging to define flammability precisely, as material behavior can differ significantly under varying conditions. Studies have shown that factors such as fiber type, twist, and the yarn count, as well as the structure of knitted fabrics, play an important role in determining ignition time and the rate at which flames spread.

The flammability, the ability to ignite easily, of the textile materials is largely influenced by the type of fiber used. Fibers can generally be categorized based on their flammability into three groups: (i) non-flammable, (ii) easily flammable, and (iii) moderately flammable. Textiles are classified into natural and synthetic polymers. Silk, cotton, flax, jute, and wool are some of the examples of natural fibres, while synthetic fibers include polyester, nylon, acrylic, aramid, and Nomex. Additionally, there exists a category known as regenerated fibers, such as viscose, rayon, and lyocell, which are man-made but derived from naturally occurring polymers. Each type of fiber exhibits distinct flammability characteristics. Fibers like cotton, acetate, triacetate, and rayon ignite easily and burn rapidly, making them highly flammable. Others, which are more difficult to ignite and may self-extinguish once removed from a flame, are considered moderately flammable. Fibers like wool, modacrylic, and aramids typically do not continue to burn once the flame source is removed and are classified as non-flammable [22].

Among the many methods of achieving flame retardancy through biotechnology, the pomegranate peel, which is rich in tannins and polyphenols, has the flame retardant property. The powder of this pomegranate peel is extracted through a standard Soxhlet method. The powders were further processed before their utilisation [36]. It is also biodegradable and nontoxic, which doesn't cause any harm to the environment, reducing the carbon footprint. The Life Cycle Assessment of microbial-based textile finishing is carried out to check its sustainability aspects. This process evaluates the environmental effects associated with utilizing microorganisms throughout the textile production process, covering every stage from the raw material acquisition to the end-of-life disposal. It involves the examination of the findings, pinpointing critical areas (hotspots) with the most significant environmental impacts, and making conclusions regarding the overall sustainability of the microbial-based finishing process [35].

5 Bio-based Dyes

Bio-based dyes are natural dyes that are synthesized by renewable or natural sources. These dyes are eco-friendly, biodegradable and have less amount of chemicals, making it the best alternative for the synthetic dyes. Some of the examples of bio based dyes are henna, rose, avocado and much more.

5.1 Natural Dye Sources

The natural dyes are the first kind of dye that one thinks of when it comes to bio-based dyes. It is the process of obtaining colorants from sources like flowers, fruits, leaves, etc. These dyes are highly sustainable and biodegradable, which gives a huge benefit to the environment. The plants play a major role in obtaining the natural colorants, as they are easily available in the environment. The colors extracted from the flowers are usually characterized by striking two or more colors together. The colors in the flowers occur due to the accumulation of pigments formed by the organs of the flowers, thus preventing the carbon emissions [23]. The crucial step for purifying the natural dyes is the natural dye extraction. It is classified into six types:

- Alkali or acid extraction
- Aqueous extraction
- Enzymatic extraction and fermentation
- Solvent extraction
- Ultrasound and microwave-assisted extraction
- Supercritical fluids [30].

The colorants obtained from nature are given in Table 1.

Table 1 Colorants obtained from nature

Material	Part	Color	Reference
Rose	Flower	Red	Guha [10]
Beetroot	Root	Red	
Madder	Wood	Red	Kothari et al. [19]
Avocado	Peel Splits	Red	
Butea (palash) monosperma	Flower	Yellow	Guha [10]
Cassia singaeara	Bark	Golden yellow	
Carrot	Roots	Orange	Kothari et al. [19]
Lily	Leaves and stalk	Green	Kothari et al. [19]
Mango	Leaves	Yellow to brown	Fauzi et al. [9]
Peltophorum	Flower	Brown or yellow	

Table 2 Colorants obtained from microbes

Microbe	Color	Reference
Carotenoids	Orange and red	[1]
Chlorophyll	Green	
Pseudomonas	Deep blue	[18]
Various	Brown to black	

5.2 *Microbial Pigments*

Other than the natural dyes, there are also many other ways one can extract colorants. The colorants obtained from microbes are given in Table 2 [12].

5.3 *Environmental Impact of Textile Recycling*

The recycling of textiles is the key point that is used to reduce the overall impact on the environment, which leads to the development of sustainable textiles. The major issue was the amount of cost that is used on the clothing and textile waste (USD 500 billion according to the past year). This information played a key role in insisting on the importance of circularity to spread information and prevent loops of each material [17]. As mentioned above, over 70% of the carbon footprint found in the textile industry is due to some processes like raw material preparation, production, and processing. While the other 30% is due to the retail and transportation of the textile industry. Studies show that it may increase by up to 60% by 2030. The Systematic Literature Review (SLR) is used to find the right amount of data to adopt into a circular economy within the textile and apparel industry. The 3 R's have also played a huge role in this. The clothing is reused without any significant change to its original form [37]. These are also regenerated cellulose obtained from the textile waste that can be used for various applications, even though they are difficult to obtain if they are not collected properly. The textile recycling waste is classified into:

- Pre-consumers
- Post-consumers
- Industrial textile waste.

Pre-consumers: This type of textile waste is known as clean waste that can be used as byproducts in the manufacturing process.

Post-consumers: This waste is usually used in towels, pillow cases, which are torn and damaged.

Industrial textile waste: These known as dirty waste and is taken away from the industry.

The circular economy managed to maintain the demand of the garments and also improve the environmental benefits of the industry (GFA and BCG).

6 Enzyme-Based Textile Processing

The role of enzymes in the textile industry has been significantly evolving for the past few years. As people prefer sustainable clothing instead of synthetic textiles. The enzymatic approach has huge advantages as the enzymes are biodegradable and don't have any chemicals in them. They also require less amount of water and can work under low temperatures unlike the traditional processes.

6.1 Role of Enzymes

As mentioned before, the enzymes are non-toxic and require less amount of energy and water. This unique characteristic can be used to replace the harsh chemicals used in the manufacturing of a textile. These enzymes can also improve the smoothness and softness of the fabric, making it durable and not needing additional processes. Amylase, laccase, and catalase are the most used enzymes in the desizing, bleaching, and degrading logging processes. Recently, cellulose has been used in the denim finishing, while the lactases are used for the decolorization of textiles. Due to all these, the enzymes now play an important role in the textile industry, reducing carbon emissions. The key applications of the enzymes fall under two main categories: The preparatory and the finishing. The various applications of enzymes will be discussed in this chapter.

6.2 Enzymes Assisted Desizing

Desizing is the process of removing the size paste from the fabric to make it more hydrophilic. Traditionally, the desizing was carried out by oxidizing agents or acids. But this led to a lot of pollution and a large carbon footprint. So the enzymes, which are usually eco-friendly, energy efficient, and generally improve the fabric quality, are used instead. An example of enzyme-assisted desizing is where a measured mixture of α-amylase and polygalacturonase (Am and PG) enzymes was used with orange peel, and their temperature, pH, and enzyme activity were observed. The cotton fabrics were effectively desized with a single bath of Am or PG [5].

6.3 Enzyme-Based Scouring

Scouring is the process of removing non-cellulosic substances such as pectin, wax, and dirt from the fabric. This stage makes the fabric suitable for bleaching. The enzyme can increase the rate and the concentration of the chemical reaction by playing the role of the catalyst [20]. It is obtained from living organisms and usually contains high molecular weight protein. In the overall 7000 enzymes, only 75 are used in the textile industry. Some enzymes require additional cofactors to make them more efficient as a catalyst. Cellulose, proteases, and pectinases are used in the bioscouring of cotton fabrics. They can be used either alone or in combination with pectinase, playing an important role [24]. These reduce the amount of carbon footprint.

- **Protease**: This enzyme is used to break down the substance and remove the unwanted proteins present in the fabric [7].
- **Lipases**: The lipases are used to increase the absorbency and dyeing properties of the fabric. It is applicable for both the bioscouring and biodesizing [13].
- **Cellulose**: Cellulose breaks down the long chains and removes the protruding fibres.
- **Amylase**: The Amylase removes the starch and desizes the lingering starch present in the fabric.

6.4 Enzymatic Bleaching

Bleaching is the process of removing the excess dirt and impurities from the fabric to make it more white/brighter. The enzymatic approach minimises the carbon footprint compared to the traditional methods. Bleaching not only increases the brightness of the fabric but also increases its strength and durability of the fabric. This method requires less amount of energy as it works at a lower temperature than other methods [2]. This enzyme acts as a catalyst and reduces the activation energy of the reaction. Denim fabric achieves a better look than a normal wash. Bio bleaching is frequently used in the denim manufacturing rather than any other material, as the lipases are specified for the indigo, a substance that gives denim its blue color. Bleaching [38] stated that the bleaching of cotton is usually done in much lower concentrations due to the laccases. The enzymatic approach is the best and most beneficial approach than any other traditional processes.

6.5 Enzyme-Based Finishes

Enzymes are essential when it comes to textile bioprocessing. Enzymes have been used around since the eighteenth century; their initial application in textile processing was less than twenty years ago. By decreasing strength and effort, enzymes enhance functionality. Chemical textile processing has existed since the first day, despite creating a risk. Instead, enzymatic processing reduces the pollution rate caused by textile processing and is non-toxic and bio-friendly. Furthermore, enzyme technology protects the most essential elements of the environment and drastically reduces the consumption of water. As a consequence, producers around the world are increasingly appreciating the advancements in enzymatic textiles. Cellulase, amylase, pectinase, catalase, laccase, and many others are the most common enzymes employed in textile processing. It also improves the quality of textiles [34].

Cellulase-based enzymatic finish on cotton

Alkaline pectinase enzymes bio-scoured cotton-based textiles, especially loomstate cotton fabric and loomstate cotton/polyester (50/50 and 35/65) blended fabrics, whereas α-amylase enzymes bio-desized them. The result of bio-scoured substrates received different procedures for bio-polishing after being bleached with peracetic acid. It depicts how conditions during treatment affect the degree of bio-polishing. The fabric loss of weight shows values that are higher than those at 40 °C and comparable at 50 and 60 °C. The whiteness index decreases by bio polishing temperatures, mainly 40° and 50°; a higher temperature, such as 60 °C, does not affect the whiteness index. In addition, it was found that mercerized cotton had higher levels of cellulase than either 100% cotton or cotton/polyester blends [4]. These treatments have significant effects on cotton-like smooth surfaces due to the removal of fuzz and fillers present on them. Enhances the appearance of the cotton fabric as polished and clearer. After finishing, the fabric gets softer and smoother. And also, light reflection will get higher in the polished cotton fabric, resulting in improved dye appearance. This finishing not only has positive effects on fabric but also the environment, as there is no need for harsh chemicals since the enzymes are biodegradable, which leads to sustainability. It also targets unwanted fiber ends without damaging the whole fabric. Unlike traditional finishing methods, which require high energy and water consumption, this method is performed at a moderate temperature, saving energy and water usage.

Using chitosan biopolymer and enzymes in the anti-felting finishing method

The high AOX (adsorbable organic halogen compounds), which causes damage to the environment, is caused by the waste of wool plants. An innovative zero-AOX processing alternative that makes use of ecologically conscious enzymes and chitosan as a biopolymer rather than traditional methods such as chlorine/Hercosett processing. To improve the shrink-proofing and anti-felt characteristics of wool, they examined oxidative, enzymatic, and additive processes along with numerous combinations of all of these techniques [28].

6.6 Sustainable Benefits of Enzyme Usage in Textiles

To extract greenhouse gases and use them in novel stabilization processes, studies have developed potent and efficient biocatalysts. In addition, these have improved enzymes to more efficiently degrade toxic and hazardous substances for biological remediation. Biocatalytic recycling is gaining momentum, with engineered cutinases and PETases developed for the depolymerization of the abundant plastic, polyethylene terephthalate (PET) [31]. Enzymes are biological catalysts that can identify a substrate and initiate synthesis and hydrolysis processes. Since enzymes can identify and react to a molecule from a pool of identical chemicals, their great specificity toward their substrates is their most important characteristic. Under some operating conditions, enzymes can become labile catalysts, which can significantly impair their stability. Nonetheless, it has been shown that attaching enzymes to solid supports is an effective way to stabilize them and maintain their catalytic capabilities [29]. By using less energy, water, and chemicals, enzymes offer significant ecological advantages in the textile processing industry. They produce less hazardous effluent, are biodegradable, and operate in mild environments. Their non-toxic nature guarantees safer working circumstances, and their targeted action maintains the quality of the cloth. Enzymes, which come from renewable resources, aid in the manufacturing of environmentally beneficial, circular textiles.

The usage of microbes in dyeing has led to tremendous environmental benefits. Colorifix, a university of Cambridge spin-off, uses microbes to produce colours and reduces the carbon footprint. As approximately 5 trillion litres of water is used for dyeing in an apparel industry, the toxins in the dyes also pollutes the water sources. Colorifix has developed a method that could develop and deposit pigments directly on the material. This biotechnological method causes much lesser harm than the conventional method. They have built a demonstrator dye house with bioreactors and fermenters. Their method improved the scalability of the dye and its cost-effectiveness. This method doesn't produce any toxic wastes. It showed that the carbon footprint can be reduced by 31% and the water by 49% [8].

7 Future Scope

Biotechnology has been rapidly evolving for the past few years and has been adapted into various industries, just like the textile industry. Biotechnology can create a promising future by reducing the carbon footprint to a minimal amount. Emerging technologies like bioengineered fibres, which can replace synthetic fibres, can give a more promising future in an eco-friendly environment. There will be many future developments in microbial dyeing and enzyme-based textile processing that will reduce the toxic chemicals to 0% and minimize the usage of water and energy. Advanced genetic engineering and mutations can also allow us to make custom-made microbes to satisfy some specific necessities. As innovations and technology develop, the textile industry is expected to adopt various sustainable practices and reduce its carbon footprint, and save the environment.

8 Conclusion

The demand for sustainable and eco-friendlier garments has been increasing rapidly over the past few years. Biotechnology allowed this achievement using its various techniques, like bio-based processes, microbial assistance, and an enzymatic approach. These innovations have revolutionized the textile industry, turning its perspective to a greener approach. This greener approach requires less energy, water, temperature, and chemicals, unlike the traditional methods. This ultimately reduces the industry's carbon footprint produced in each manufacturing process. The sustainability of the industry also increases, allowing the younger and future generations to enjoy these benefits. This chapter insists on a sustainable and eco-friendly approach to each process using biotechnology-based materials to reduce the industrial carbon footprint and to sustain the health of the environment.

References

1. Abdelfattah A, Ali SS, Ramadan H, El-Aswar EI, Eltawab R, Ho SH, Elsamahy T, Li S, El-Sheekh MM, Schagerl M, Kornaros M (2023) Microalgae-based wastewater treatment: mechanisms, challenges, recent advances, and future prospects. Environ Sci Ecotechnology 13:100205
2. Jadhav A (2007) Bio-processing of textile. (Online). Available from https://www.fibre2fashion.com/industry-article/2477/bio-processing-of-textiles. Accessed on 17 May 2025
3. Afraz N, Uddin F, Syed U, Mahmood A (2019) Antimicrobial finishes for textiles. Curr Trends Fashion Technol Text Eng 4(5):87–94
4. Ali H, Hashem M, Shaker N, Ramadan M, El-Sadek B, Hady MA (2012) Cellulase enzyme in bio-finishing of cotton-based fabrics: effects of process parameters. Res J Text Appar 16(3):57–65
5. Aly AS, Sayed SM, Zahran MK (2010) One-step process for enzymatic desizing and bioscouring of cotton fabrics. J Nat Fibers 7(2):71–92
6. Anaheim CA (2026) Innovative eco-friendly textile dyeing methods. (Online). Available from https://www.sustainablemanufacturingexpo.com/en/articles/innovative-eco-friendly-dyeing.html. Accessed on 17 May 2025
7. Bristi U, Pias AK, Lavlu FH (2019) A sustainable process by bio-scouring for cotton knitted fabric suitable for next generation. J Text Eng Fash Technol 5:41–48
8. Colorifix (2020) A revolutionary dyeing process to help the textile industry reduce its environmental impact. (Online). Available from https://cordis.europa.eu/article/id/442393-natural-dyeing-process-delivers-environmental-benefits. Accessed on 18 Aug 2025
9. Fauzi AM, Sunarti TC, Yulianto A (2024) Prospects of the development of the sustainable natural textile dye industry: A Systematic Literature Review. In: IOP Conference Series: Earth and Environmental Science, 1359(1):012078. IOP Publishing
10. Guha AK (2019) A review on sources and application of natural dyes in textiles. Int J Text Sci 8(2):38–40

11. Gupta B, Revagade N, Hilborn J (2007) Poly (lactic acid) fiber: an overview. Prog Polym Sci 32(4):455–482
12. Gurkok S, Ozdal M (2023) Microbial pigments in biotechnology. Biotechnol Action Unveiling Nat Potential 7–26
13. Hasan KF, Rahman MM, Rima FK, Sultana J, Taher MA, Horváth PG, Garab J, Bejó L, Alpár T (2025) Sustainable prospects of lignocellulosic wood and natural fiber-based materials in 3D and 4D printing. Adv Compos Hybrid Mater 8(2):189
14. Hökl L (2023) Closing the loop: exploring the circular economy potential of cotton and the role of biodegradation as a sustainable end-of-life solution. Doctoral dissertation, University of Innsbruck
15. Ibrahim NA, Gouda M, Husseiny SM, El-Gamal AR, Mahrous F (2009) UV-protecting and antibacterial finishing of cotton knits. J Appl Polym Sci 112(6):3589–3596
16. Islam MT, Asaduzzaman S (2019) Environmentally-friendly textile finishing. Text Clothing 101–129
17. Juanga-Labayen JP, Labayen IV, Yuan Q (2022) A review on textile recycling practices and challenges. Textiles 2(1):174–188
18. Kadam S, Deore S, Harshal T, Wagh V, Thube U (2024) The use of biological pigments in cosmetics for eco-friendly and sustainable coloring. Int J Pharm Qual Assur 15:539–545. https://doi.org/10.25258/ijpqa.15.1.81
19. Kothari S, Khan S (2021) Natural dyes using plant palette: a brief review. J Glob Biosci 10(4):8582–8605
20. Kushwaha A, Kesarwani P, Kushwaha R (2024) Bio-scouring-an advancement in preliminary processing of textile. Int J Creative Res Thoughts 12:340–345
21. Madej-Kiełbik L, Gzyra-Jagieła K, Jóźwik-Pruska J, Wiśniewskia-Wrona M, Dymel M (2022) Biodegradable nonwoven materials with antipathogenic layer. Environments 9(7):79
22. Mazumder NUS, Islam MT (2021) Flame retardant finish for textile fibers. Innov Emerg Technol Text Dye Finish 373–405
23. Melo Miranda B, Vilela Junior O, Santos Fernandes S, Mendes Lemos GR, Schwan CL, Aliaño-González MJ, Fernández Barbero G, Murowaniecki Otero D (2025) Potential of new plant sources as raw materials for obtaining natural pigments/dyes. Agronomy 15(2):405
24. Mojsov K (2014) Trends in bio-processing of textiles. Adv Technol 3(2):135–138
25. Haripriyaa M, Suthindhiran K (2024) Waste management using marine microorganisms. Sustain Environ 10(1):2324574
26. Natarajan G, Rajan TP, Das S (2022) Application of sustainable textile finishing using natural biomolecules. J Nat Fibers 19(11):4350–4367
27. Nofal RM (2022) Biodegradable textiles, recycling, and sustainability achievement. In: Handbook of biodegradable materials, pp 1–37
28. Onar N, Sariişik M (2004) Application of enzymes and chitosan biopolymer to the antifelting finishing process. J Appl Polym Sci 93(6):2903–2908
29. Ottone C, Romero O, Urrutia P, Bernal C, Illanes A, Wilson L (2021) Enzyme biocatalysis and sustainability. Nanostructured catalysts for environmental applications. Springer International Publishing, Cham, pp 383–413
30. Pizzicato B, Pacifico S, Cayuela D, Mijas G, Riba-Moliner M (2023) Advancements in sustainable natural dyes for textile applications: a review. Molecules 28(16):5954
31. Radley E, Davidson J, Foster J, Obexer R, Bell EL, Green AP (2023) Engineering enzymes for environmental sustainability. Angew Chem Int Ed 62(52):e202309305
32. Radhakrishnan S (2014) Application of biotechnology in the processing of textile fabrics. In: Roadmap to sustainable textiles and clothing: eco-friendly raw materials, technologies, and processing methods. Springer Singapore, Singapore, pp 277–325
33. Rungruangkitkrai N, Phromphen P, Chartvivatpornchai N, Srisa A, Laorenza Y, Wongphan P, Harnkarnsujarit N (2024) Water repellent coating in textile, paper and bioplastic polymers: a comprehensive review. Polymers 16(19):2790
34. Sarkar S, Soren K, Chakraborty P, Bandopadhyay R (2020). Application of enzymes in textile functional finishing. In: Advances in functional finishing of textiles. Springer Singapore, Singapore, pp 115–127

35. Subramanian K, Chopra SS, Cakin E, Li X, Lin CSK (2020) Environmental life cycle assessment of textile bio-recycling–valorizing cotton-polyester textile waste to pet fiber and glucose syrup. Resour Conserv Recycl 161:104989
36. Thirunavukarasu B (2025) Enhancing flame retardant and self-cleaning properties of cotton fabric with bio natural material and TiO_2 nanoparticles
37. Thomas K, Durrani H, Brady J, Ludwig K, Yatvitskiy M, Clarke-Sather AR, Cao H, Cobb K (2024) Fundamental challenges and opportunities for textile circularity. Sustainability 16(24):11117
38. Tzanov T, Basto C, Gübitz GM, Cavaco-Paulo A (2003) Laccases to improve the whiteness in a conventional bleaching of cotton. Macromol Mater Eng 288(10):807–810
39. Wen OY, Tohir MZM, Yeaw TCS, Razak MA, Zainuddin HS, Hamid MRA (2023) Fire-resistant and flame-retardant surface finishing of polymers and textiles: a state-of-the-art review. Prog Org Coat 175:107330

Technological Innovations and Climate Adaptation Strategies for Road Management

Victor Hugo Souza de Abreu

Abstract This chapter investigates the impacts of climate change on road transport infrastructure and operations, emphasizing the sector's vulnerability to extreme weather events and long-term climate variability. Adopting a qualitative and exploratory approach based on a narrative literature review, the study identifies and categorizes major climate-related disruptions that affect freight and passenger transportation systems, including physical damage, operational interruptions, safety risks, and logistical inefficiencies. A matrix of climate impacts and associated technological responses was developed to synthesize the findings. The study highlights a range of technological solutions that can enhance the resilience of road transport, such as intelligent transportation systems, climate and traffic monitoring technologies, low-emission vehicles, adaptive infrastructure materials, and logistics management tools. These solutions were further examined in relation to contemporary frameworks, including Nature-Based Solutions (NbS), resilient city initiatives, and systemic risk management approaches. Additionally, the research aligns its analysis with the Sustainable Development Goals (SDGs), emphasizing how climate adaptation in transport contributes to global sustainability agendas.

Keywords Climate change adaptation · Road transport infrastructure · Technological innovation · Transport resilience · Intelligent Transportation Systems (ITS) · Climate risk management

---

V. H. S. de Abreu (✉)
Laboratory of Innovation in Transportation and Energy Resources (LITEN), Civil Engineering Course, Polytechnic School, Federal University of Rio de Janeiro (UFRJ), Rio de Janeiro, Brazil
e-mail: vhsa@poli.ufrj.br

S. S. Muthu (ed.), *Reduction of Industrial Carbon Footprint*, Environmental Footprints and Eco-design of Products and Processes,
https://doi.org/10.1007/978-3-032-10913-2_7

1 Introduction

The road transport sector is highly affected by the impacts (short, medium, and long term) of climate change and will continue to face significant challenges such as reduced efficiency of transport operations and infrastructure capacity to withstand extreme weather events outside the "typical" range [19, 7–9]. Transportation networks play a key role in the economy and society [30, 13, 26], making it essential to adapt these infrastructures to climate change [3, 4, 31, 10].

Decision-makers in the public and private transport sectors responsible for infrastructure development and improvement face a crucial new challenge in the face of the growing issue of climate change [7, 14–17]. This issue is considered one of the greatest global challenges facing the planet today [31], causing global economic losses of more than US$ 500 billion [1]. The impacts of climate change on the transportation network can lead to the loss of infrastructure assets and hinder the recovery and resilience of the entire system [32].

Considering that transportation infrastructure represents a significant national investment, it is essential to address these challenges to ensure its adequate protection and adaptation [5]. Thus, the development and adoption of new technologies can play a key role in minimizing these impacts [21].

This study aims to map the impacts of climate change on the road transport sector, covering both freight and passenger transport, through a literature review. In addition, it seeks to identify and analyze the available technologies that can be implemented to increase the resilience of this sector in the face of climate challenges. It is worth noting that special attention is given to investigating the potential of these strategies to achieve the Sustainable Development Goals (SDGs) of the United Nations (UN) 2030 Agenda, focusing on those most directly related to the theme of road transport and climate resilience.

It is believed that, through the literature review applied in this study, it was possible to identify the different ways in which climate change affects road transport and understand the best practices to make this sector more resilient through available technologies, such as intelligent traffic management systems, electric and hybrid vehicles, climate monitoring and adverse conditions forecasting systems, among others.

By mapping the impacts of climate change and identifying promising technologies, this study sought to provide valuable information for decision-makers in the road transport sector. Based on these insights, they will be able to take proactive measures to adapt and strengthen infrastructure and operations, thereby reducing risks and increasing the sector's resilience in the face of ever-evolving climate challenges.

2 Methodology

This study adopts a qualitative, exploratory, and descriptive approach, based on a narrative literature review, to identify the main impacts of climate change on road transport infrastructure and mapping technological and strategic solutions capable of increasing the sector's resilience. The research did not follow formal systematic review protocols such as PRISMA or meta-analysis. Instead, it employed a structured bibliographic survey strategy guided by thematic relevance, the timeliness of publications, and the methodological consistency of the sources consulted.

The review encompassed both national and international scientific and technical literature, including peer-reviewed journal articles, institutional reports (such as those by the IPCC), technical documents from government agencies, and proceedings from specialized conferences on transport, infrastructure, and climate change. Preference was given to studies published in the last ten years, particularly those addressing the relationship between extreme climate events, vulnerabilities in road infrastructure, and technological adaptation proposals.

Following data collection, the content was analyzed using a thematic extraction strategy, identifying recurring climate impacts on road operation, safety, logistics, and physical infrastructure. These impacts were categorized according to their direct and indirect effects and organized into an impact-solution matrix, which links each identified impact to a set of technological or operational measures discussed or applied in specialized literature.

The technological solutions were grouped according to their practical application, including weather monitoring and early warning systems, vehicle-embedded technologies, resilient construction materials, intelligent transportation systems (ITS), alternative energy sources, and logistics management strategies. Where possible, the study highlighted the practical applicability of these solutions based on case studies, integrative reviews, and documented evidence from different national and international contexts.

Beyond the technological dimension, the study incorporated a complementary analysis grounded in contemporary frameworks such as Nature-Based Solutions (NbS) and Resilient Cities, aiming to discuss the integration between grey and green infrastructure as a systemic and multifunctional strategy to face climate challenges. The study also considered how this adaptation strategies align with the United Nations SDGs, to frame the research within a broader global context of sustainability and climate governance.

This methodological approach enabled the construction of a critical and multidimensional overview of the climate challenges affecting the road transport sector and the potential strategies to address them. Although it does not constitute a formal systematic review, the analysis was conducted with interpretive rigor and conceptual coherence, providing a robust foundation to support planning decisions, public policy development, and private-sector strategies for climate adaptation in road transport systems.

3 Results and Discussion

To develop context-appropriate adaptation measures, as presented in Table 1, the first step is to identify the locations where climate threats and the most intense biophysical impacts occur. The effective allocation of resources to potential projects based on this information is one of the crucial responsibilities of infrastructure planners [2].

According to the IPCC [20], Climate Risk Analysis should cover three main components: climate threat, exposure, and vulnerability, which includes sensitivity and adaptive capacity. Climate threat refers to the probability of occurrence and intensity of extreme weather events or gradual changes in climate (such as rising average temperatures or sea level rise). Exposure refers to the location and extent of assets or populations subject to these threats. Vulnerability represents the degree of susceptibility of these systems to climate impacts and is determined by two main factors: sensitivity (i.e., the magnitude of the negative response to the impact) and adaptive capacity (the ability to anticipate, resist, absorb, and recover from adverse effects).

This analysis makes it possible to identify critical points in the road infrastructure, characterize the predominant risks in each region, and thereby provide a technical basis for choosing the most effective, efficient, and cost-effective solutions to increase the climate resilience of the transportation system. This analysis should make use of tools that assist decision-makers through observational and/or experimental data, as well as emission scenario simulation models, to quantify the impacts of extreme events and incremental climate change [6, 11]. By doing so, it becomes possible to identify critical points in the road network in various geographical locations around the world.

In addition to conducting a comprehensive Climate Risk Analysis, it is equally essential to apply cost-effectiveness assessments to select and prioritize technological adaptation measures. Such analyses allow for the comparison of different intervention alternatives based on the relationship between the costs of implementing, operating, and maintaining the solutions and the expected benefits in terms of damage reduction, operational continuity, and extension of infrastructure life.

Cost-effectiveness analysis is particularly important in contexts of scarce resources and multiple competing demands for public and private investment. It provides technical and economic support so that decision-makers can choose strategies that are not only environmentally and technically viable, but also economically viable, socially beneficial, and financially resilient over time.

Thus, the adoption of evidence-based adaptation measures that meet economic efficiency criteria becomes a key element in ensuring the sustainability of adaptation policies and the rational allocation of resources, reducing future risks and promoting greater safety and functionality in road transport in the face of climate change [8].

The technological solutions outlined in Table 1 can help address these challenges, from the use of more resistant construction materials and monitoring systems to extend the useful life of roads and bridges, to the use of electric vehicles, more efficient logistics systems, and renewable energy sources to reduce greenhouse gas

Table 1 Mapping the impacts of climate change on the transport sector and how to minimize them

Impacts	How to minimize?
Disruption of road transport routes due to gravitational displacement of boulders and/or rocks	(1) Implementing weather monitoring systems can help anticipate extreme weather events and plan alternative routes in advance, minimizing transportation disruption. (2) Vehicles equipped with satellite navigation systems can use real-time information to navigate alternative routes, minimizing time lost in traffic jams. (3) The implementation of smart infrastructure, such as smart traffic lights, traffic sensors, and vehicle-to-infrastructure (V2I) communication technology, can help better manage traffic flow and minimize congestion in the event of disruptions to transportation routes
Increased risk of accidents on roads and highways due to extreme events that cause biophysical impacts such as flooding and gravitational displacement of soil and/or rock	(1) The use of early warning systems can help provide real-time information on adverse weather conditions and potential road hazards, enabling them to make more informed decisions about alternative routes, delivery times, and necessary resources. (2) The use of vehicles equipped with advanced safety technology, such as automatic braking and stability control systems, can help minimize the risk of accidents. (3) The use of electric vehicles can help minimize the risk of accidents caused by adverse weather conditions, as these vehicles tend to have better traction and control on wet or snowy roads compared to fossil fuel-powered vehicles. (4) The implementation of smart infrastructure, such as road sensors and V2I systems, can help detect adverse conditions and alert drivers to potential hazards on the roads
Negative impacts on the seasonality of goods production and distribution	(1) The use of demand forecasting systems can help road freight companies anticipate fluctuations in transport demand during different periods of the year, allowing them to adjust their operations accordingly. (2) The use of inventory management systems can help adjust inventory levels to meet fluctuations in demand for goods during different periods of the year, reducing the need for additional transportation. (3) The use of intermodal transportation, which combines different modes of transportation (e.g., road, rail, and sea), can help reduce dependence on road transportation and minimize the impact of climate change on transportation demand
Negative impacts on production and freight transport, causing disruptions in the supply chain	(1) The use of weather monitoring systems can help predict extreme weather events in advance, allowing transport and logistics companies to plan alternative routes or postpone production and transport of goods to avoid disruptions in the supply chain. (2) Storing goods in strategic locations can help minimize the impact of supply chain disruptions caused by extreme weather events, allowing companies to continue to deliver their products to customers during periods of disruption. (3) Building adaptive infrastructure can help minimize damage caused by extreme weather events such as floods and droughts by making roads and highways more resilient and adaptable to adverse weather conditions. (4) The use of rail and maritime transport can help minimize the impact of extreme weather events on the transport of goods, as these modes of transport are less susceptible to disruptions caused by extreme weather events than road transport

(continued)

Table 1 (continued)

Impacts	How to minimize?
Sea level rise and coastal erosion can affect coastal road transport routes, forcing transport companies to find longer alternative routes and increasing operating costs	(1) The construction of coastal barriers can help protect coastal road transport routes from the effects of sea level rise and coastal erosion. These barriers can be made of natural materials, such as sand dunes, or artificial materials, such as concrete walls. (2) Elevating roads and highways, as well as elevated bridges and tunnels, can help minimize the impact of sea level rise and coastal erosion on coastal road transport routes, making them more resistant to the effects of extreme weather. This technology may require significant investment, but it can help transportation and logistics companies maintain efficient and safe operations. (3) The use of sensors can help monitor the condition of coastal roads in real time, allowing transportation companies to avoid routes affected by sea level rise and coastal erosion and reduce response times to these events
Rising temperatures can affect the efficiency of internal combustion engines, reducing their effectiveness and increasing fuel consumption	(1) An efficient cooling system can help maintain the engine at appropriate temperatures, allowing it to operate at maximum efficiency. This can also help ensure that drivers feel more comfortable in high-temperature environments. (2) A more radical solution would be to switch from internal combustion engines to more efficient engines that generate less heat and use less fuel. This may involve the use of more advanced technologies, such as electric or hybrid engines. (3) The use of more heat-resistant materials can help protect engine components that are exposed to high temperatures, allowing them to function more efficiently and last longer. This may include the use of composite materials, ceramics, or other advanced materials
Exposure to climate risks increases the need for insurance for road vehicles and raises the operating costs of transportation companies	(1) Installing weather sensors in vehicles can help monitor weather conditions and alert drivers to potential hazards, allowing them to adjust their driving and avoid accidents. These sensors can also be used to collect climate data that can be used to predict future risks and help plan safer routes. (2) Telematics is a technology that enables real-time collection of vehicle data, including information on driver behavior, road conditions, and weather conditions. This data can be used to identify patterns and emerging risks, allowing transportation companies to take preventive measures to reduce exposure to weather risks. (3) Remote monitoring allows transportation companies to monitor their vehicles in real time using cameras and other sensors. This can help identify problems early and prevent damage caused by extreme weather events

(continued)

Table 1 (continued)

Impacts	How to minimize?
In some cases, load restrictions may be necessary to protect road infrastructure damaged by extreme weather events	(1) Load monitoring technology can help ensure that weight limits are enforced. Sensors installed on the vehicle can measure the load and alert the driver if it exceeds the permitted limit. This can help avoid fines for overloading and reduce the risk of damage to road infrastructure. (2) Transportation management systems can help transportation companies plan routes and loads more efficiently, taking into account load restrictions imposed on certain routes. This can help reduce the number of trips required and minimize operating costs. (3) Road infrastructure damage detection technology, such as vibration or deformation sensors, can be used to identify damaged areas on a road. This can help determine where load restrictions are needed and prevent overweight vehicles from passing through these areas
Strong winds pose a significant threat to road transport safety and operations, especially in open areas, on bridges, elevated viaducts, and exposed logistics corridors	(1) Adoption of real-time anemometric monitoring systems integrated into road infrastructure, capable of detecting sudden wind variations and issuing alerts to operators and drivers; (2) Installation of variable message signs (VMS) that inform about adverse conditions and recommend speed reduction or alternative routes; (3) Creation of aerodynamic barriers installed in critical sections, designed to reduce wind intensity on the road. (4) In addition, embedded technologies, such as vehicle stability sensors and lane correction assistance, also contribute to minimizing the effects of crosswinds on commercial vehicles
Climate change may lead to increased regulatory pressure on road transport companies to reduce GHG emissions and adopt sustainable transport practices	(1) The adoption of electric or hybrid vehicles can help significantly reduce greenhouse gas emissions from road vehicles. In addition, battery technology has improved greatly, allowing these vehicles to have greater autonomy. (2) Biofuels, such as biodiesel and ethanol, can be used as a substitute for fossil fuels and help reduce greenhouse gas emissions. (3) Fuel management systems allow transportation companies to monitor and control the fuel consumption of their vehicles. This can help reduce greenhouse gas emissions and operating costs. (4) Intelligent transportation systems (ITS) can help improve the efficiency of freight transportation by reducing the need for unnecessary trips and increasing vehicle utilization. This can help reduce greenhouse gas emissions and operating costs. [1], for example, indicate that intelligent transportation systems can reduce carbon dioxide emissions by approximately 60%

(continued)

Table 1 (continued)

Impacts	How to minimize?
Prolonged exposure to high temperatures and heavy rainfall can reduce the lifespan of roads and bridges, increasing the need for frequent repairs and delaying the transport of goods	(1) The use of more resistant construction materials, such as high-strength concrete, can help extend the useful life of roads and bridges, reducing the need for frequent repairs. (2) Sensors and monitoring systems can be used to monitor the condition of roads and bridges, identifying problems early before they become more serious. This can help reduce the need for repairs and delays in the transport of goods. (3) Permeable pavements allow water to pass through the pavement rather than accumulate on the road surface. This can help reduce the effects of heavy rainfall, extending the life of the road and reducing the need for frequent repairs. (4) Drainage systems can be used to help drain water from roads and bridges, reducing water accumulation and extending the life of roads and bridges. (5) Reflective paints and coating can be used to help reduce the temperature of roads and bridges, reducing the effects of high temperatures and extending the life of roads and bridges

Source Love et al. [23]; Rattanachot et al. [29], Moretti and Loprencipe [25]; de Abreu et al. [4], de Abreu et al. [6], Krichen et al. [21]

emissions. In addition, more advanced construction techniques, weather forecasting systems, and solutions to improve road safety can also help make road transport more sustainable and resilient to the impacts of climate change.

3.1 Aligning Technological Solutions with Nature-Based Solutions

It should be noted that the technological solutions presented in Table 1 need to be linked to NbS through the provision of ecosystem services, which can generate numerous benefits for people and address multiple challenges simultaneously [24]. These solutions cover a spectrum of approaches, such as the integration of natural elements into road infrastructure and the restoration of adjacent ecosystems [12].

These integrated solutions are essential for addressing the complex socio-ecological issues that characterize contemporary urban environments [18]. By combining innovative technologies with strategies that value natural processes—such as ecosystem restoration and green infrastructure—and participatory social actions, these measures promote multiple co-benefits [22].

Among these benefits are improved air and water quality, increased urban biodiversity, reduced heat islands, and a stronger social fabric through community engagement. In addition, these integrated strategies significantly increase the resilience of cities, preparing them to respond and recover more effectively from the adverse impacts of climate change, extreme events, and anthropogenic pressures [22].

Therefore, investing in adaptive solutions that reconcile technological, natural, and social aspects not only contributes to environmental sustainability but also strengthens cities' ability to promote the social and economic well-being of their populations, ensuring the continuity of essential services and quality of life in a context of increasing vulnerability.

3.2 Alignment of Technological Solutions with Resilient Cities Projects

It should also be noted that the efficient management of natural resources, combined with resilient cities projects, can be significantly enhanced by the use of artificial intelligence (AI) technologies [1, 28]. The application of AI allows for real-time monitoring of environmental variables, more accurate prediction of extreme events, and optimization of urban systems such as transportation, energy, water supply, and waste management.

As a result, decisions become more informed and dynamic, promoting environmental, economic, and social sustainability. In particular, the integration of spatial, climate, and infrastructure data, processed by intelligent algorithms, contributes to the construction of urban environments that are more adaptable to climate change, efficient in the use of resources, and safer for the population [28].

3.3 Special Attention to the Cascading Effects of Critical Infrastructure

Another aspect that deserves special mention is the vulnerability of critical infrastructure, such as transportation systems, to the impacts of extreme weather events. When these infrastructures are affected, a series of cascading effects can be triggered that reverberate on other essential networks, such as electricity, water supply, and telecommunications, significantly increasing the fragility of urban systems [27].

This complex interconnection requires an integrated and multidisciplinary approach to urban risk management that considers the theory of cascading, interconnected, and compound risks, as proposed in recent literature. In this context, the challenge is not only to prepare for isolated events, but also to understand and manage the systemic nature of risks, which manifest themselves at multiple levels and domains, often simultaneously [27].

Therefore, a consistent approach that articulates emergency management strategies, organizational resilience, and adaptation to climate change is essential. In this sense, the originality and value of this approach lie in proposing a paradigm that transcends the traditional fragmentation between emergency management, organizational resilience, and climate adaptation, promoting a unified concept of operational continuity for the mitigation of systemic risks. This integration represents a crucial step forward in strengthening the capacity of urban systems to resist, recover, and adapt to the challenges posed by climate change and the growing complexity of interconnected infrastructures [27].

4 Contributions to Sustainable Development Goals

The search for technological solutions that increase the resilience of road transport in the face of climate change is in line with the United Nations 2030 Agenda, particularly with regard to promoting resilient infrastructure, climate action, and building safer, more sustainable, and inclusive cities. The proposals presented in this study contribute directly and indirectly to the achievement of several Sustainable Development Goals (SDGs), with emphasis on the following:

- SDG 8—Decent Work and Economic Growth: Modernizing transportation infrastructure based on sustainable and adaptive technologies can stimulate regional development and the creation of skilled jobs in engineering, information technology, and logistics services. This contributes to target 8.2 ("progressively increase labor productivity through diversification, technological upgrading, and innovation").
- SDG 9—Industry, Innovation, and Infrastructure: This study contributes to target 9.1 ("develop quality, reliable, sustainable, and resilient infrastructure, including regional and cross-border infrastructure") by proposing the adoption of smart infrastructure, highly durable materials, climate monitoring systems, and vehicle-embedded technologies. The incorporation of these solutions stimulates the modernization of the transport sector, reduces the vulnerability of the road network, and promotes innovation as a structural axis of climate adaptation.
- SDG 11—Sustainable Cities and Communities: By addressing the effects of climate change on mobility and transport infrastructure, the study reinforces the importance of safer, more resilient, and more accessible urban and interurban systems, aligning with target 11. 2 ("provide access to safe, affordable, accessible, and sustainable transport systems for all") and target 11.b ("substantially increase the number of cities and human settlements adopting and implementing integrated policies and plans toward inclusive, resource-efficient, climate-resilient, and climate-friendly development"). Reducing accident risks, ensuring operational continuity, and integrating clean technologies into transportation help build communities that are better prepared for extreme events.

- SDG 12—Responsible Consumption and Production: The use of technologies that extend the useful life of infrastructure, reduce operational losses, and improve transportation efficiency contributes to target 12.2 ("achieve sustainable management and efficient use of natural resources"). By encouraging the use of permeable pavements, sustainable materials, and optimized drainage systems, the study promotes practices that minimize the environmental impact of highway construction and maintenance.
- SDG 13—Action against Global Climate Change: The core of this study is aligned with SDG 13, especially target 13.1 ("strengthen resilience and capacity to adapt to climate-related hazards and natural disasters") and target 13.2 ("integrate climate change measures into national policies, strategies, and planning"). The application of Climate Risk Analysis, combined with the identification of adaptive technological solutions, represents a structured and targeted approach to reducing climate vulnerabilities in one of the country's most polluting and susceptible sectors.
- SDG 17—Partnerships and Means of Implementation: Strengthening partnerships between governments, the private sector, academia, and civil society is essential for the effective implementation of technological solutions that increase the resilience of road transport. This cooperation promotes knowledge transfer, the sharing of financial resources, and joint innovation, contributing to target 17.6 ("significantly increase North–South, South-South, and triangular cooperation in science, technology, and innovation") and target 17. 17 ("encourage and promote effective public–private, public-public, and civil society partnerships"). The integration of these partnerships strengthens governance, expands access to international climate finance, and promotes institutional capacity building to address climate challenges in a coordinated and sustainable manner.

By integrating the SDGs into climate adaptation strategies in the transportation sector, this study offers valuable tools for public managers, policymakers, infrastructure planners, and logistics operators. The direct link to the SDGs allows technical decisions to be aligned with global sustainability commitments, enabling access to international climate finance, strengthening sector governance, and promoting synergies between environmental, economic, and social agendas.

In addition, the approach presented can support the private sector in defining ESG (Environmental, Social and Governance) strategies, providing indicators and pathways to meet corporate goals for decarbonization, climate risk reduction, and social and environmental responsibility.

5 Final Considerations

The main objective of this study was to identify the main impacts of climate change on road transport and to map technologies capable of increasing the resilience of this sector. Through a systematic literature review and the organization of a matrix of impacts and technological solutions, it was possible to achieve this objective by presenting a comprehensive diagnosis of the climate challenges faced by road infrastructure, as well as concrete alternatives for mitigation and adaptation.

The results showed that climate impacts affect not only the physical integrity of infrastructure, but also road safety, logistical efficiency, operating costs, and the environmental performance of operations. In response to these challenges, it was possible to identify several technological solutions—such as smart sensors, low-emission vehicles, forecasting and management systems, and resilient construction materials—that can be applied in an integrated manner to strengthen the transportation system in the face of extreme events and incremental climate change.

The analysis also highlighted the importance of conducting a Climate Risk Analysis in advance, involving the dimensions of threat, exposure, and vulnerability, as well as cost-effectiveness assessments to support decision-making based on technical, economic, and social criteria. These elements are fundamental to guide investments in adaptation and ensure that available resources are applied strategically, maximizing benefits and minimizing future risks.

Given this scenario, the findings of this study provide relevant insights for decision-makers, public managers, urban planners, and companies in the transportation sector, contributing to the development of public policies, investment plans, and corporate strategies aligned with the climate agenda and the principles of sustainability.

Although this study adopted a qualitative, exploratory, and descriptive approach based on a narrative literature review, it is important to acknowledge certain methodological limitations that may affect the scope and generalizability of the results. The choice not to follow formal systematic review protocols, such as PRISMA or to conduct meta-analyses, introduces a degree of subjectivity in the selection and interpretation of sources. Furthermore, the analysis did not incorporate quantitative modeling, simulations, or empirical field evaluations, which limits the ability to precisely quantify climate impacts and the effectiveness of technological solutions. These constraints, however, do not undermine the relevance of the proposed diagnosis, but rather highlight the need for future research involving empirical studies, regional comparative analyses, and cost–benefit assessments of the recommended measures.

As a follow-up, it is recommended that future research advance in quantifying the costs and benefits associated with each type of technology mapped, as well as in modeling regional climate scenarios applied to the transportation sector. In addition, empirical case studies in different geographical contexts can contribute to validating the effectiveness of the proposed solutions and generating practical evidence that strengthens the sector's adaptive capacity in the face of the challenges posed by climate change.

References

1. Chen L, Chen Z, Zhang Y, Liu Y, Osman AI, Farghali M, Hua J, Alfatesh A, Ihara I, Rooney D, Yap PS (2023) Artificial intelligence-based solutions for climate change: a review. Environ Chem Lett 21(5):2525–2557
2. Chinowsky PS, Price JC, Neumann JE (2013) Assessment of climate change adaptation costs for the US road network. Glob Environ Change 23(4):764–773. https://doi.org/10.1016/j.gloenvcha.2013.03.004
3. de Abreu VHS, Ribeiro FB, Santos AS (2021a) Impactos Da Mudança Climática Na Infraestrutura De Transporte Rodoviário À Base De Revisão Bibliográfica. In: 35º ANPET—Congresso de Pesquisa e Ensino em Transportes
4. de Abreu VHS, Ribeiro FB, Santos AS (2021b) Impactos da Mudança Climática na Infraestrutura de Transporte Terrestre à Luz de Revisão Bibliográfica com Abordagem Bibliométrica. In book: Via Viva 2021. Publisher: Ministério da Infraestrutura. https://doi.org/10.54507/viaviva.2021.01
5. de Abreu VHS, Ribeiro FB, Santos AS (2022) Identificação de Indicadores de Vulnerabilidade para Análise de Risco Climático da Infraestrutura Rodoviária. In book: Via Viva 2022. Publisher: Ministério da Infraestrutura
6. de Abreu VHS, Santos AS, Monteiro TGM (2022) Climate change impacts on the road transport infrastructure: a systematic review on adaptation measures. Sustainability 14(14):8864. https://doi.org/10.3390/su14148864
7. de Abreu VHS et al (2022) Identificação De Ameaças E Impactos Da Mudança Climática Na Infraestrutura De Transporte Rodoviário. MIX Sustentável 8(3):142–156. https://doi.org/10.29183/2447-3073.MIX2022.v8.n3.142-156
8. de Abreu VHS et al (2023a) Climate change adaptation strategies for road transportation infrastructure: a systematic review on flooding events. Transp Syst Technol Integr Manag 5–30. https://doi.org/10.1007/978-981-99-1517-0_2
9. de Abreu VHS, de Lima GBC, Monteiro TGM, Santos AS (2023b) Implications of climate change for the Brazilian road infrastructure. In: COVID-19 and climate change in BRICS nations. Routledge, pp 80–99
10. de Abreu VHS, Oda S, Monteiro TGM (2025) Strategic decisions: a support model for prioritizing projects in the Brazilian transport infrastructure. MIX Sustentável 11(1):107–118
11. de Abreu VHS, dos Santos RF, Oda S, Santos AS (2024) Análise de riscos climáticos em infraestruturas de transportes: uma exploração metodológica. Boletim do Observatório Ambiental Alberto Ribeiro Lamego 18(1):28–44
12. de Abreu VHS, dos Santos RF, Oda S, Santos AS (2024b) Nature-based solutions to reduce carbon footprint and increase resilience of road transportation infrastructure. In: Carbon footprint assessments: case studies & best practices. Springer Nature Switzerland, Cham, pp 159–176
13. de Abreu VHS (2024) Implementing circular economy principles in road transport: enhancing efficiency to mitigate resource depletion and carbon footprint. In: Carbon footprint assessments: case studies & best practices. Springer Nature Switzerland, Cham, pp 289–312
14. de Assis TF, de Abreu VHS, Marujo LG, de Almeida D'Agosto M (2024) Best practices applied in logistic intern: From the perspective of reducing the carbon footprint. In: Carbon footprint assessments: case studies & best practices. Springer Nature Switzerland, Cham, pp 313–336
15. de Assis TF, Monteiro TGM, de Abreu VHS, D'Agosto MDA, Santos AS (2022a). Enabling the green bonds market for sustainable transport projects based on the measure/monitoring, reporting and verification method. In: Carbon footprints of manufacturing and transportation industries. Springer Nature Singapore, Singapore, pp 1–24
16. de Assis TF, Ricci LM, Monteiro TGM, de Abreu VHS, D'Agosto MDA, Santos AS. (2022b) Sustainable transport indicators and mitigation actions applied to the green bond principles. In: Carbon footprints of manufacturing and transportation industries. Springer Nature Singapore, Singapore, pp 139–169

17. da Costa MG, de Abreu VHS, de Assis TF, da Costa VX, de Almeida D'Agosto M, Santos AS (2022) Life cycle assessment and circular economy strategies for electric vehicle: a systematic review on mitigating climate change and reducing resource depletion in road transportation. In: Carbon footprints of manufacturing and transportation industries, pp 113–137
18. dos Santos RF, de Abreu VHS, Oda S, Santos AS (2024) Climate response and nature-based solutions: evaluating the impact of the belt and road initiative in Latin America. In: Carbon footprint assessments: case studies & best practices. Springer Nature Switzerland, Cham, pp 273–288
19. Evans C, Tsolakis D, Naudé C (2009) Framework to address the climate change impacts on road infrastructure assets and operations. In: ATRF Conference. Disponível em: https://www.australasiantransportresearchforum.org.au/sites/default/files/2009_Evans_Tsolakis_Naude.pdf. Acesso em: 01 maio 2023
20. Intergovernmental Panel on Climate Change—IPCC. Climate Change 2014: Impacts, adaptation, and vulnerability. Part A: global and sectoral aspects. In: Field CB, Barros VR, Dokken DJ, Mach KJ, Mastrandrea MD, Bilir TE, Chatterjee M, Ebi KL, Estrada YO, Genova RC, Girma B, Kissel ES, Levy AN, MacCracken S, Mastrandrea PR, White LL (eds) Contribution of working group IIto the fifth assessment report of the intergovernmental panel on climate change. Cambridge University Press, Cambridge, United Kingdom and New York, NY, USA, 1132p
21. Krichen M, Abdalzaher MS, Elwekeil M, Fouda MM (2024) Managing natural disasters: an analysis of technological advancements, opportunities, and challenges. Internet Things Cyber Phys Syst 4:99–109
22. Lin BB, Ossola A, Alberti M, Andersson E, Bai X, Dobbs C et al (2021) Integrating solutions to adapt cities for climate change. Lancet Planet Health 5(7):e479–e486
23. Love G, Soares A, Püempel H (2010) Climate change, climate variability and transportation. Procedia Environ Sci 1:130–145. https://doi.org/10.1016/j.proenv.2010.09.010
24. McPhearson T, Cook EM, Berbés-Blázquez M, Cheng C, Grimm NB, Andersson E et al (2022) A social-ecological-technological systems framework for urban ecosystem services. One Earth 5(5):505–518
25. Moretti L, Loprencipe G (2018) Climate change and transport infrastructures: state of the art. Sustainability 10(11):4098. https://doi.org/10.3390/su10114098
26. Oda S, Abreu VHSD, Fernandez SDDS, Sá TSW (2024) Utilização de rejeitos de mineração na construção de pavimentos rodoviários. Matéria (Rio de Janeiro) 29(2):e20240142. https://doi.org/10.1590/1517-7076-RMAT-2024-0142
27. Pescaroli G, Guida K, Reynolds J, Pulwarty RS, Linkov I, Alexander DE (2023) Managing systemic risk in emergency management, organizational resilience and climate change adaptation. Disaster Prev Manag 32(1):234–251
28. Rane N, Choudhary S, Rane J (2024) Artificial intelligence for enhancing resilience. J Appl Artif Intell 5(2):1–33
29. Rattanachot W et al (2015) Adaptation strategies of transport infrastructures to global climate change. Transp Policy 41:159–166. https://doi.org/10.1016/j.tranpol.2015.03.001
30. Santos AS et al (2021) An overview on costs of shifting to sustainable road transport: a challenge for cities worldwide. In: Carbon footprint case studies: municipal solid waste management, sustainable road transport and carbon sequestration, pp 93–121. https://doi.org/10.1007/978-981-15-9577-6_4
31. Santos AS, Ribeiro SK, de Abreu VHS (2020) Addressing climate change in Brazil: is Rio de Janeiro city acting on adaptation strategies? In: 2020 International conference and utility exhibition on energy, environment and climate change (ICUE). IEEE, pp 1–11. https://doi.org/10.1109/ICUE49301.202
32. Trindade DQL, Abreu VHS, Santos AS (2019) Medidas de Adaptação frente aos Possíveis Impactos das Mudanças Climáticas no Sistema de Transportes: Uma Revisão da Literatura. In: XXXIII Congresso Nacional de Pesquisa e Ensino em Transportes

MIX
Papier aus verantwortungsvollen Quellen
Paper from responsible sources
FSC® C105338

If you have any concerns about our products,
you can contact us on
ProductSafety@springernature.com

In case Publisher is established outside the EU,
the EU authorized representative is:
Springer Nature Customer Service Center GmbH
Europaplatz 3, 69115 Heidelberg, Germany

Printed by Libri Plureos GmbH
in Hamburg, Germany